13.46

LONGMAN KEY SKILLS

LEVEL

1+2

INFORMATION TECHNOLOGY

Series editor: Barry Smith

Longman

Longman Key Skills
titles available in the series

Application of Number Level 1+2
Application of Number Level 3

Communication Level 1+2
Communication Level 3

Information Technology Level 1+2
Information Technology Level 3

Pearson Education Limited
Edinburgh Gate, Harlow
Essex CM20 2JE, England
and Associated Companies throughout the world

First published 2000

British Library Cataloguing in Publication Data
A catalogue entry for this title is available from the British Library

ISBN 0-582-42488-7

Set by 3 in Sabon and Quay Sans
Printed in Great Britain by Henry Ling Ltd,
at the Dorset Press, Dorchester

Contents

How to use this book

This book helps you obtain the key skill called Information Technology level 1 or level 2. You will be doing your key skills with your other studies in a school, college or at work. The common combinations are:

Level 1
GCSE and key skills
GNVQ Part One and key skills
GNVQ Foundation and key skills
NVQ1 and key skills

Level 2
GCSE and key skills
GNVQ Part One and key skills
GNVQ Intermediate and key skills
NVQ2 and key skills

An Information Technology key skill is not asking you to be a computer expert. It is about knowing some facts about modern information systems and communication technology and showing that you can actually use them in real life. Most of us need a reminder about how to use IT and this book is organised to provide rapid help when you want it.

The good news about gaining any of the key skills is that you don't always need to do extra work. The evidence for the key skill is produced while you are doing your normal study and work such as in the classroom, laboratory, workshop, or while working at a job.

Of course there is a certain cunning in knowing which of your work to keep and how to show it, and that's what this book is about. There are special sections for all popular GCSE and GNVQ subjects which tell you exactly what you need to do.

You can use this book in different ways; it depends on what you need. For example, you might not need to read it from the beginning. To get the most out of this book, have a look at the following summary of how it is organised and decide how you can use it best.

The GNVQ Advanced awards are now called **Vocational A-levels**.

From September 2001 GNVQ Foundation and Intermediate awards are likely to be known as **Vocational GCSEs**.

Part 1: The Learning Curve

This part of the book concentrates on what you need to know to get the key skill units. It has useful information about computers and the Internet, and how to use word processing, database, spreadsheet and graphics programs. It concentrates on the more tricky ideas and has clear worked examples to show you how to use them.

You can check that you have the basic knowledge needed by the key skill units. If you are up to speed with your IT then you may not need much of this section.

Part 2: The Bottom Line

This part of the book tells you what you must do to gain the key skills units. It explains:

- The words and ideas of the key skills
- The difference between level 1 and level 2
- How you can practise the skills
- What must be in your portfolio of evidence

Your collection of evidence or portfolio is the key to getting your key skill. This part of the book tells you how to choose your evidence and get it ready.

Part 3: Opportunities

This part of the book tells you where to find opportunities for evidence in the study or work you are already doing. If you are at school or college, you should look up the pages for your particular subjects at GCSE or GNVQ.

Everyone can make a start on using IT to find information by looking at the chapter on **Information sources**. It lists some useful and interesting website addresses.

Margin

Look in the margin for simple explanations of important words and ideas and for references to other places in the book where there is useful information.

Part 1: The Learning Curve

This part concentrates on what you need to know to get your key skills qualification. It will show you:

- How to use computers and the Internet in practical situations.
- Techniques for getting the most out of word processing, database, spreadsheet and graphics programs.
- Clear explanations of the more difficult ideas.

This part is divided into nine sections:

- **Introduction to IT and ICT**
- **Using computers**
- **Using databases**
- **Using the Internet**
- **Calculations**
- **Using text**
- **Using graphics**
- **Communicating**
- **Glossary**

Introduction to IT and ICT

IT is shorthand for information technology. The alternate abbreviation ICT is shorthand for information and communications technology.

Computers are an important part of IT but there are many other forms of information technology. You are also using IT when you use a phone with memory buttons, when you program your video recorder, withdraw money from an automatic cash dispenser, or shop at a store with a barcode reader.

This part concentrates on areas of IT where people often need help, especially computers. The content is suitable for beginners. It is not technical but you may have to learn some new ideas.

Some of the ideas have their own special words and it is useful to know them. These words are often placed in the margin, alongside a relevant section. At any time you can also look up words in the glossary on page 38.

Examples of IT/ICT
computers
email
webpages
mobile phones
phone services
CD-ROMs
digital TV
teletext
videoconferencing
cash dispensers
smart tills

How IT can help you

Information technology helps our lives in many ways; here are some examples:

- New types of job, e.g. working at home
- New types of business, e.g. business on the Internet
- New ways of getting information, e.g. travel timetables
- New ways of sending and receiving messages
- New ways of shopping
- New ways of paying, e.g. electronic money

You need to be able to say what you would have done before you could use IT.

Benefits and disadvantages of IT

Using IT makes our lives easier but some of the advantages go hand in hand with possible disadvantages. Some of the issues are described in the table and you should at least know that there are pros and cons. When thinking about advantages and disadvantages of IT you should always:

- Keep an open mind about benefits and drawbacks
- Concentrate on facts not opinion

LEVEL 2

IT effect	Some advantages	Some disadvantages
Job changes	New jobs are created Existing jobs are more interesting People can work from home without travel	Loss of old jobs Possible health hazards from using IT equipment Loss of contact with people at work
Electronic money (e.g. swipe cards)	Security from theft Quicker transfer between accounts	You need a bank account Not suitable for small items There may be charges
IT in shops (e.g. electronic tills and stock control)	Faster turnover at checkouts Better flow of goods to the shelves	Possible stress on shop assistants Difficult for small shops
E-commerce (e.g. shopping on the Internet)	You can order from anywhere in the world Prices are competitive	Someone still has to deliver goods to your home
Communications	Rapid email to anywhere in the world Targeting mail to particular names and addresses	Loss of privacy
Private life	Easier to get information Easier shopping Cheaper quotes for travel and other sevices Possible new jobs	Loss of privacy Feelings of increased pressure

Using computers

This chapter tells you, or reminds you, about the basics of using your computer:

- The parts of a computer
- Using Windows to drive your computer
- Making changes to your work
- Saving your work and finding it again
- Printing out your work

Parts of a computer

Hardware

processor (CPU)
memory (RAM)
disk storage
screen (VDU)
keyboard
mouse, joystick
printer, plotter

A modern computer is a machine which automatically does set things by following a set of instructions called a **program**. Usually we just need to know that a computer program does useful tasks for us, such as word processing and that we can 'drive' the program.

A large computer is called a **mainframe** but you will normally use the computer found in homes and offices called a **personal computer** (PC). All computers have parts which are either hardware or software:

- **Hardware** is the computer equipment.
- **Software** is the set of programs which tell the computer what to do, such as run a game or be a word processor.

The computer reads and uses software programs stored on magnetic disks, on CD-ROM or in other computers connected by networks.

Software

Windows program
word processor
programs
CAD programs
games programs

Using Windows

When working with computers you need an interface so that you can put information in and get information back. The common types of interface on a personal computer are:

- **Input:** keyboard, mouse
- **Output:** screen, printer

You operate the inputs and outputs with the help of a program such as **Windows.** The opening screen of Windows shows little pictures, called icons, which will start a program or open a document.

TYPES OF COMPUTER DISK

- Hard disks are sealed inside the computer.
- Floppy disks can be removed by the user.
- CD-ROMs can also be removed by the user.

Note: The spelling is **disk** not **disc**.

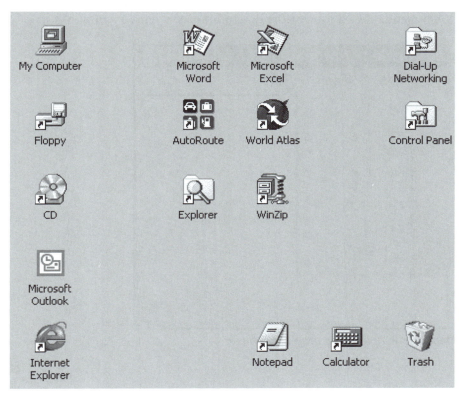

Typical icons on a computer 'desktop'

Selecting

The most common way of making a choice on a computer screen is to use a mouse or a similar rolling device attached to the computer. When you roll the mouse, you see a **cursor** move over the screen. Some computers have rollerballs, small sticks or sensitive pads which can also move the cursor over the screen.

You move the cursor over the screen until it rests on the icon or words that you want to **select**. You complete the selection by one of the following methods:

- Point with the mouse and click or double-click on the mouse button.
- Point with the roller ball or pad and double-click on special buttons.
- Point to the icon with a special stick and tap the screen.
- Press the **Enter** key on the keyboard.

An **icon** is a small picture used instead of words; clicking on an icon usually starts an activity.

Screen Menus

Your choices of actions for a program are usually given by a screen **menu**. The place to find a menu is in the **menu bar** at the top of the screen. Programs have their commands grouped in a similar pattern arranged in the following order:

File   Edit   View   Insert   Format   Help

Pointing and clicking on one of these commands will produce a drop-down **menu**; have a look at the diagram to see what a menu looks like.

The **File** group of commands is always on the left.
The **Help** group of commands is always on the right.

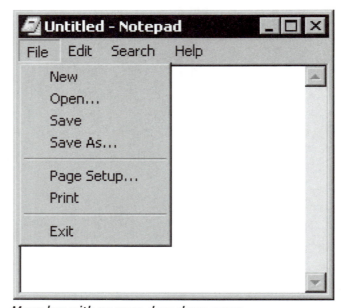

Menu bar with an open drop-down menu

Making changes

Highlight your choice then . . .

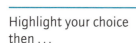 **Cut**

 Copy

 Paste

All programs allow you to make changes and this process is often called editing. Clicking on the **Edit** group of the menu bar at the top of the screen gives a drop-down menu with editing options available for the program you are in. All programs allow you to make the following changes:

- **Cut** deletes but keeps an invisible copy.
- **Copy** leaves the original untouched but makes a copy.
- **Paste** produces a copy at the cursor position.

You choose the text or image to be edited by selecting it:

- **Text:** hold down the mouse button while you paint text.
- **Pictures:** move the mouse onto the image and click or double-click

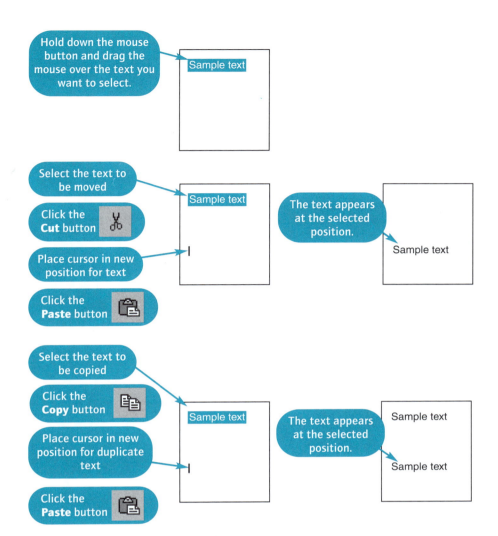

Editing text using Cut, Paste and Copy

Saving and finding work

You have spent hours at the computer, your work looks wonderful on-screen, but it will all be lost once you switch the computer off. So, when you have produced something on a computer, you need to save it somewhere on the computer. And when you have saved your work, you need to be able to find it again.

Folders and files

Filing cabinets are boring but, when organised, we can quickly find stuff we have filed away. Computers may store your work on electronic disks but they are organised like cabinets with folders and files. Folders may also be called directories.

From IT helpdesk 'top ten'
I did save my work but I don't know where to find it!

- **File:** one document such as a letter, a report, a set of calculations.
- **File name:** your choice of name for a file.
- **Folder:** a collection of files on the same topic such as one subject you are studying, your emails or your hobbies.

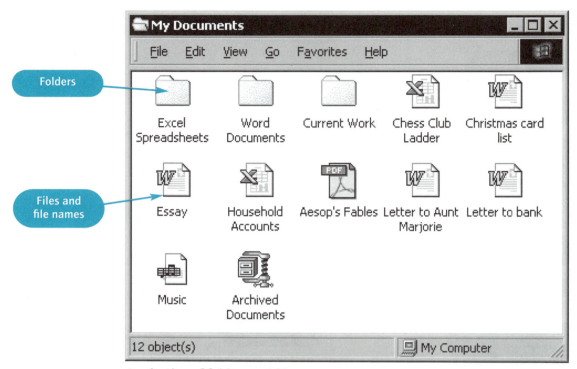

A selection of folders and files

Saving

Your computer may suggest you store your files in a folder called something like **My Documents** and this is a good start. When working on networks of computers, such as at college or school, the folders will already be in place for you. But on your own computer you should set up a personal system of folders with your own labels.

> ### LOST AND FOUND
>
> Can't find a folder or file? Click on the Windows **Start** button, usually in the bottom left corner. Then click on the **Find** icon which pops up. Enter details in the boxes to start the finding process.

The first time you save your work, you will be asked for a file name. This is a label that you must think up and enter. At first you may be tempted by very short file names but, after losing your work and your memory a few times, you should use names which make reasonable sense. Earlier computer systems don't allow names longer than eight letters. Suppose you want to save a new document for the first time:

- At the top left of the screen click on **File**.
- Move down the menu and click on **Save as**.
- Click on the entry box labelled **Save in**.
- Select a folder to store the file in.
- Click on the entry box labelled **File name**.
- Type in your choice of name for the file.
- Click on the **Save** button.

The second time you save your work, the computer will remember the file name and use it. Get into a routine of saving your work on the following occasions:

- Every five minutes
- Every new page
- When interrupted by someone
- When the phone rings
- When the cat approaches
- When you leave the computer, even for a moment

That way you won't have lost too much if you make a mistake, or the power fails.

The Learning Curve

Old saying
Save often –
it costs nothing

> **BEWARE**
>
> - Programs also have files which sit in folders with labels like **Windows**.
> - Do not save your personal files in program folders.
> - Do not alter any program folders or files – it usually stops the program working.

Printing

Everything on the computer screen can be printed onto paper. We usually use a word processor or drawing package to produce paper copies. Before you print you should check the following points:

- Printer connected to computer?
- Printer switched on?
- Paper loaded?
- No paper jams?

Before you print onto paper you can see exactly how it will look by using **Print preview**. The toolbar has an icon for this or you can pull down the **File** menu.

Options for printing

The program needs to match the page on your screen to the paper in your printer. Check this by clicking on **File** then **Page setup** and review the following options.

Options	Effect
Margins	Allows you to change the blank spaces between the edges of the paper and your work; the standard settings are usually good
Paper size	Allows you to change the size of paper; A4 paper is the usual choice
Orientation	Allows you to print across the paper; normal printing is down the paper (portrait) but for some projects you might print across the paper (landscape)

Portrait   Landscape

When you are ready to print, choose which pages and how many copies.

Options	Effect
Number of copies	The standard setting is to print one copy, but it can be any number
All pages	Will print all pages of your document, including any blanks at the end
Current page	Will print just the page which is on-screen when you start the **Print** command
Some pages	Will print just the pages that you put in the box or boxes, e.g. 3-6 will print pages 3, 4, 5 and 6

Using databases

We often need to store information, to sort it and then find it. The folders in a filing cabinet are a form of database. But the use of computers has made it easier to store large amounts of information and to search through it quickly.

Someone still has to design and organise the database and to keep it tidy. To make the best use of the information in a database, you should know a few rules and tricks described in this section.

Uses of databases

A database is used whenever you need to do the following types of job:

- Holding lots of information, e.g. the details of everyone who has car insurance with a company
- Sorting files into a different order, e.g. listing in order of name or type of car
- Searching all information and producing lists, e.g. listing all people who need to renew insurance this month
- Getting statistics, e.g. counting how many cars are of a certain make or calculating their average value

You don't have to have a huge amount of information to use a database. A good example of a simple database is a collection of names and addresses for people and you might already keep them in address books or on card files. The computer programs for simple databases often have names like Address Book or Cardfile.

A computer database of information needs to be organised with a structure. It is useful to think of a collection of information on cards when dealing with the items of a database:

- **A field** contains one item of information about a person, such as the name, town or phone number.
- **A record** contains all the information for one person, such as on one card in the file.
- **The database** is the full collection of records.

Information retrieval is about getting the special information you want from a large store of information.

Examples of databases
card index of library books
electronic address book
school timetable
choices of course at college
airline timetable

Some home databases
personal numbers stored in phones favourite TV channels

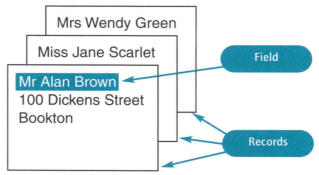

Parts of a database

Entering information

When you buy a ready-made database, like the address list in an email program, it will already have a structure. You just need to enter the information, with one record for each person. The makers of the software will have chosen a structure which suits most people. There are often options to change the structure, perhaps by adding another field, e.g. a field for an extra phone number.

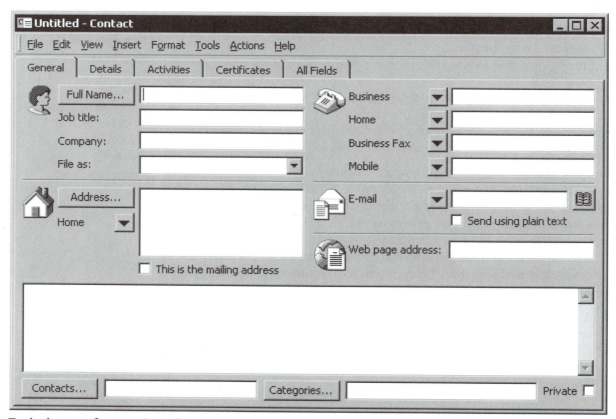

Typical screen for entering information into a database

Look at the diagram that shows database records of 10 people, together with their towns and their ages. This information is laid out in the form of a grid or table. It is the raw information of a database before any sorting of filtering. Even in this simple database there are two different types of data:

- **Text:** letters of the alphabet such as names, and some numbers used for labels such as phone numbers.
- **Number:** numbers used for calculation, such as finding the average age.

Some data types
text (alphanumeric)
number (numeric)
date

The Learning Curve

Field   Record

Forename	Surname	Town	Age
James	Goodman	Northtown	16
Elizabeth	Baker	Northtown	18
Michael	King	Southport	15
Neville	Madden	Southport	19
Kirit	Patel	Northtown	16
Michael	Schofield	Westville	21
Mary	King	Northtown	16
Paulo	Dias	Westville	16
Nita	Bhudia	Southport	17
Peter	Zhang	Northtown	20

Sample database

The complete information about a single person makes up one **record** of data. The records can be entered in any order, providing that you put each item of information in its correct **field**. In this database the fields are forename, surname, town and age. The power of the database will allow us to sort it then search it.

Output from a database

To use a database we normally ask for some output or result. This result might be a particular record we want, or it might be a sorted list of names. Most databases have ready-made commands which let you sort and find simple information. Here are some typical outputs from the example database.

All data, sorted in order of surname

Forename	Surname	Town	Age
Elizabeth	Baker	Northtown	18
Nita	Bhudia	Southport	17
Paulo	Dias	Westville	16
James	Goodman	Northtown	16
Michael	King	Southport	15
Mary	King	Northtown	16
Neville	Madden	Southport	19
Kirit	Patel	Northtown	16
Michael	Schofield	Westville	21
Peter	Zhang	Northtown	20

All data, sorted in order of age

Forename	Surname	Town	Age
Michael	King	Southport	15
Paulo	Dias	Westville	16
James	Goodman	Northtown	16
Mary	King	Northtown	16
Kirit	Patel	Northtown	16
Nita	Bhudia	Southport	17
Elizabeth	Baker	Northtown	18
Neville	Madden	Southport	19
Peter	Zhang	Northtown	20
Michael	Schofield	Westville	21

Data filtered to match the town of Southport

Forename	Surname	Town	Age
Michael	King	Southport	15
Nita	Bhudia	Southport	17
Neville	Madden	Southport	

Selection of different outputs from the sample database

Searching a database

See also: **Searching the Internet**, page 16.

To use a database you need to be able to look through all the records and find what you want:

- **A query** is a search of a database with a set of conditions.
- **A match** is when the computer finds a record which agrees with your conditions.
- **A report** is a list of all the records which obey the conditions.

Suppose you want to find all records in a database that have the surname Goodman. This is your query or search condition. Different databases can ask you to enter your condition in different ways. Some examples are shown in the following table.

LEVEL 2

Query methods	Action	Note
Entry box with label such as **Find What?**	Enter **Goodman** in box	Simple databases of addresses use this method
A query table which allows you to copy an example on-screen	Enter **Goodman** in table under surname label	Popular databases use this method
Entry line for you to enter a mathematical condition	Enter **Town=Southport**	Lists all records where the town is Southport
	Enter **Age >16**	Lists all records where the age is greater than 16
	Enter **Town=Southport AND Age <16**	Lists only those records where the town is Southport and the age is greater than 16

Seach shorthand

a = b	a equals b
a > b	a is greater than b
a < b	a is less than b
a<>b	a not equal to b

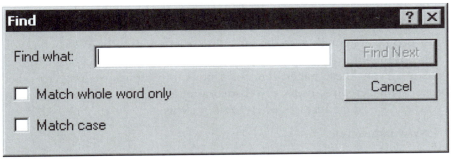

Simple search dialog box

You often get a chance to sharpen your search as you enter your conditions into the database search window. Most software packages on your computer have a simple search mechanism. Pull down the **Edit** menu and try the **Find** command. Near the entry box there are often other options. The following table gives some outcomes for 'disc'. Fortunately, most simple searches aren't fussy about capital letters, unless you tell them to be fussy.

Uppercase means capital letters.
Lowercase means small letters.

Possible extra search condition	Will find
Ignore case (ignore capital letters)	disc, Disc, DISC
Whole words only	disc but not discovery
Find sound-a-likes	disc, disk

Using the Internet

When computers are connected together they make a network, such as you might get in a school or college. On a larger scale you can have a wide area network (WAN), such as when cash dispensers in the streets are linked to large computers in the banks.

The Internet is a system which allows any computer in the world to join together a system of link-ups. At home we usually link to the Internet via our phone line but larger computers in the Internet are connected by high-speed telecommunication links which use cable and satellite.

The Internet can be used in different ways but the two main uses for most people are **email**, described in a later section, and the **Web**. The Web is part of the Internet where organisations or people have pages of text and pictures. From these pages you find things, buy things or jump to other pages. You get directly to a **website** by entering the address into your browser. Website addresses often look like this:

www.name.com

www.name.co.uk

Networks

local area network (LAN)

wide area network (WAN)

World Wide Web (WWW)

Looking up websites

To use the Internet you subscribe to an **internet service provider** (ISP) who has a computer which links you to the Web. You connect your computer to your ISP using a modem and a special phone number. Your computer will have an icon for dial-up connections. Click on this icon to connect to the Internet; you may need to enter a password before the modem begins to dial. At work, college or school your computers may be permanently connected to the Internet.

When you're online to the Web you will use a **browser**, such as **Microsoft Explorer** or **Netscape Navigator**. The browser is a computer program which displays webpages.

Using a browser

All types of browser work in a similar manner and have a banner along the top of the screen where you click to give commands. The main commands are explained in the following table.

Internet activities

send and receive email
check football scores
update news
browse catalogues
book holidays
shop and buy
buy and sell at auctions
check TV and films
book cinema tickets
visit museums
play/download music
play games
visit fan clubs
read books
chat with friends
check your bank

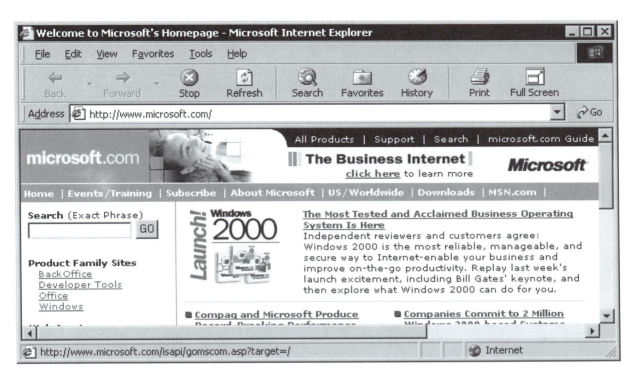

Internet browsers: Internet Explorer and Netscape

Some ISPs
AOL
BT Internet
Bun
Claranet
Compuserve
Freeserve
Virgin Net

Icon	Effect when clicked
Address or Go to	Allows you to enter the address or name of a website that you want
Back	Takes you back to the webpage you just visited
Forward	Moves you forward to a webpage you just came from
Stop	Stops loading the current page
Refresh or Reload	Loads a new version of the current page; useful when a page is incomplete or often updated
Home	Goes to the page seen when your browser opens; this can be changed to whatever you want
Search	Begins options which search by keyword
Favourites or Bookmarks	Drops down a list of favourite websites you have marked in the past
Mail	Connects to email
History	Gives a list of the websites you have visited in past days or weeks
Mail	Opens the options for using email
Print	Prints out the webpage shown on-screen

Most general website addresses start with **www**.

The most important thing is to get started using the blank line near the top of the screen called **Address** or **Go to**. Suppose you want to see the website with address www.bbb.co.uk:

- Type it into the address box of your browser.
- Check you have typed it correctly.
- Press the **Enter** key
- Wait for the webpage to download into your computer.

> **STOP BUTTON**
>
> If a download is slow because it has many images, click the **Stop** button once the text is on-screen. If you want to see an image then right-click on the blank image. Also, use the **Stop** button if a connection is very slow; you can try again at a less busy time.

Searching the Internet

The Internet joins you to the information held by thousands of computers worldwide, so it is the largest database in the world. Somewhere on the Internet are websites with answers to your questions, but they have to be found. You can reach this information in the following two ways:

- **Go directly** because you know the website address.
- **Use a search engine** to trawl the net for relevant websites.

LEVEL 2

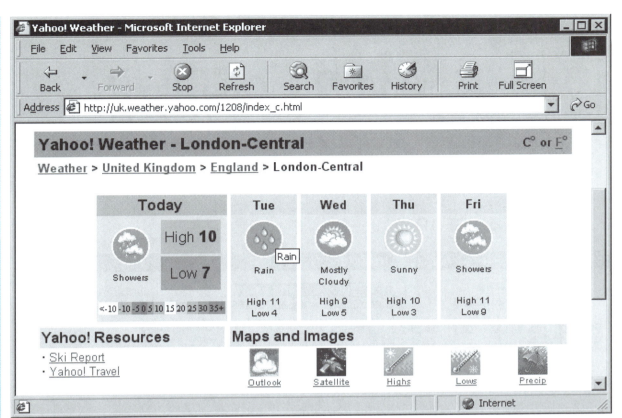

Typical Web page

Search engines

Some well-known search engines are described in the following table. Some of the engines let you enter your question in general language. Others work better if you use the tricks for advanced searches.

Search engine and address	Notes
AskJeeves **www.ask.co.uk**	Good site for beginners. Allows you to ask questions in plain English such as, Where is information about xyz? Some of the answers may be too general
MSN Search **www.search.msn.com**	Good site for beginners and good links to other information
Yahoo **www.yahoo.co.uk**	Can be browsed by categories or can be searched by keyword
AltaVista **www.altavista.com**	Indexes more webpages than many engines. Also has advanced search option to focus your seach

Popular search engines
AltaVista
AskJeeves
Excite
Google
HotBot
Infoseek
Lycos
MSN Search
Northern Light
Yahoo

Two typical search engines (AskJeeves and AltaVista)

Advanced searches

Search shorthand

a = b	first equals the second
a > b	a is greater than b
a < b	a is less than be

Some types of search involve the logic of combinations. This is particularly true when searching huge databases like the Internet. For example, searching for 'spice girls' could bring you thousands of references for 'spice' as in cooking and 'girls' as in girls.

But if you want to find the pop group Spice Girls then you can link the two words together by using AND. This will force the search engine to look for references where the words are joined together. Here is how you search for websites about blue whales.

Extra search conditions (word or symbol)	Possible effect
AND +	Will find references which include the word *blue* joined with the word *whales*. This condition will probably find the references to the particular type of whale
OR —	Will find references which include either the word *blue* or the word *whales*. This condition will find thousands of references you don't want

Calculations

Spreadsheets

A spreadsheet is a computer program which can do the sort of calculations that repeat themselves in a boring way.

Spreadsheets look like a big sheet of paper with a grid of boxes or a table. The clever thing is that each box, or cell, contains its own calculator and word processor, and these cells can be linked together to do useful things. Changing just one number on-screen causes all other numbers to be updated automatically.

Graphs
All spreadsheets can turn your numbers into smart graphs.

> Spreadsheets are good places to store numerical information because they can sort the numbers and turn them into graphs if you like.

A spreadsheet is included in most collections of computer software; here are some possible uses:

- Calculator
- Quick graphs
- Estimates of cost
- Invoices
- Profit and loss
- Business plans
- Models and graphs
- Mathematics

Some spreadsheets
Excel
1-2-3
Ability
Works

All spreadsheet programs work in the same general way and have the same basic layout. In any cell you can enter the following types of information:

- **Text:** usually labels such as Sales or January.
- **Numbers:** usually data such as sales figures.
- **Formulas:** such as A1+A2 which will add the contents of cell A1 and cell A2.

Above the basic worksheet of cells on-screen there are standard menus and icons where you can give commands to the spreadsheet. For example, there may be icons which make it easy to add numbers, sort columns of data and produce graphs.

When you change any number on-screen the formulas will automatically recalculate all other numbers and show the new results. That's the point of a spreadsheet.

Parts of a spreadsheet

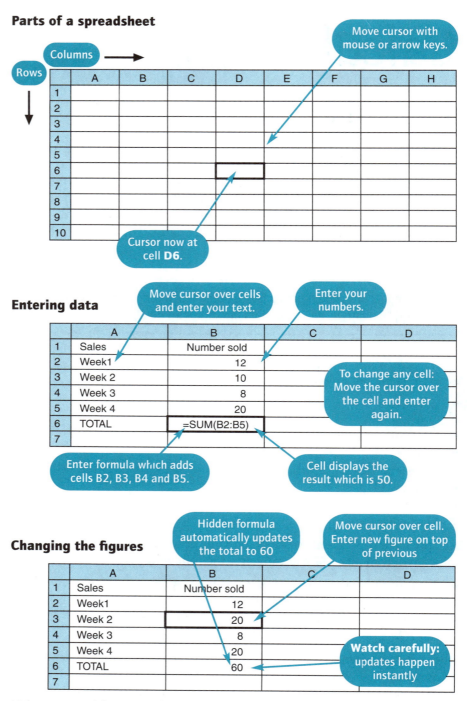

Move cursor with mouse or arrow keys.

Columns →

Rows

	A	B	C	D	E	F	G	H
1								
2								
3								
4								
5								
6								
7								
8								
9								
10								

Cursor now at cell **D6**.

Entering data

Move cursor over cells and enter your text.

Enter your numbers.

	A	B	C	D
1	Sales	Number sold		
2	Week1	12		
3	Week 2	10		
4	Week 3	8		
5	Week 4	20		
6	TOTAL	=SUM(B2:B5)		
7				

To change any cell: Move the cursor over the cell and enter again.

Enter formula which adds cells B2, B3, B4 and B5.

Cell displays the result which is 50.

Changing the figures

Hidden formula automatically updates the total to 60

Move cursor over cell. Enter new figure on top of previous

	A	B	C	D
1	Sales	Number sold		
2	Week1	12		
3	Week 2	20		
4	Week 3	8		
5	Week 4	20		
6	TOTAL	60		
7				

Watch carefully: updates happen instantly

Using a spreadsheet to calculate the sum of a column of figures

Spreadsheet formulas

A spreadsheet has more maths power than most people can ever use. But you should at least learn the useful features such as automatically adding a set of numbers or producing an average. Here are some common symbols used for spreadsheet maths.

Entry	Typical example	Effect
Cell reference	D6	Refers to the cell in column D at row 6
Range	D6:D10	All the cells in column D between row 6 and row 10
=		Enter = to start all maths in spreadsheets
+	= 3+3 = A1+A2	Addition
−	= 8−3 = A1−A2	Subtraction
*	= 8*3 = 8*D6	Multiplication
/	= 8/3 = D6/A1	Division
%	= 100+17.5%	Percentage
^	= 10^3	Exponent or power, e.g. $10^3 = 1000$
SUM	= SUM (A1:A22)	Adds all numbers in the cells between A1 and A22
AVERAGE	= AVERAGE (A1:A22)	Gives the mean of all numbers in the cells between A1 and A22
>	= A1>B1	Greater than (nl) Tells you whether it is true or false
<	= A1<B1	Less than (nl) Tells you whether it is true, or false

Changing the look of spreadsheets

All spreadsheets can have their appearance changed; this may be useful for printing out and showing results to other people. You can use the commands of your particular spreadsheet to change the following features:

- Column widths
- Changing text to bold and other effects
- Position of text in cell (left, centre, right)
- Position of number in cell (left, centre, right)
- Making gaps between rows and columns
- Moving cells or rows and columns
- Copying cells or rows and columns

Charts and graphs

All spreadsheets allow you to convert a set of figures into a graph or a choice of graphs. It is worth entering a set of your figures into a spreadsheet just to get a good graph. Here are some of the possibilities.

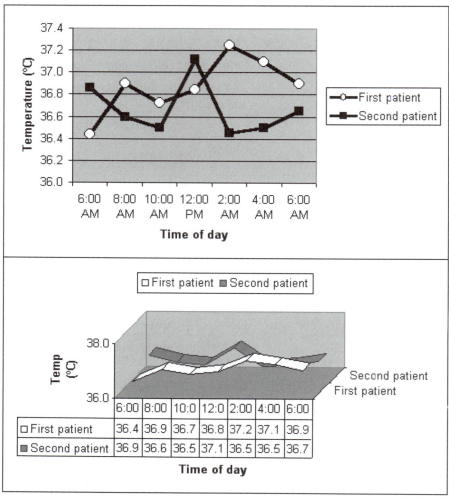

Two graphs created within a spreadsheet (both graphs use the same information)

Suppose you want to make a chart or graph out of some numbers, perhaps the link between people's weight and height.

Entering the data

- Start a new spreadsheet which is blank.
- Enter the names for labels into separate cells, e.g. Weight, Height.
- Enter the values for (weight, height) pairs into separate cells.

Weight (kg)	61	63	64	65	66	66	67	67
Height (cm)	152	157	165	165	170	175	177	182

Making the graph

- Select the labels and the numbers that you want to see as a graph.
- Use the **Tools** menu or the chart icon to start the chart making.
- Use the options to choose the type of graph.
- Use the options to point to the labels and to add headings.
- Use the **View** menu to preview the effect.
- You can print the chart or import it into another program such as your word processor.

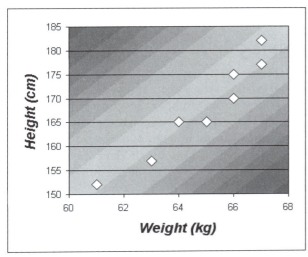

Your graph might look like this!

Using text

Word processors

Word processors
Word
SmartSuite
WordPerfect
Works
Ability Office
WordPad

Word processors (WPs) let you write on the screen and change your words or change the layout before you print onto paper. So you really are processing words. You can change the size and style of the printing on the page and add items such as photos or other images. Most word processors also help you make clever layouts such as columns, bullets and tables.

A word processor is included in most collections of computer software; here are some possible uses:

- Letters and notes
- Reports and essays
- Notices and newsletters
- Forms and tables
- Mailshot letters

All word processors work in the same general way and have the following basic features:

Some WP features
generate text (writing)
change text (editing)
change layout of text
change look of print
automatically label
    pages
automatically number
    pages
insert pictures

- **Writing text** allows you to enter your work at the keyboard.
- **File commands** allow work to be saved, opened, previewed and printed.
- **Edit commands** allow work to be changed using undo, cut, paste and copy.
- **View commands** allow changes to the look of the word processor on-screen.
- **Insert commands** allow pictures and other effects to be brought in from other programs or files.
- **Format commands** allow changes to the type style and page layout.
- **Headers and footers** show a label across the top and bottom of each page; this can include the page number.

WYSIWYG

The normal operation of a word processor is WYSIWYG (what you see is what you get), which means that the styles and layout you see on-screen will be the same on the printout.

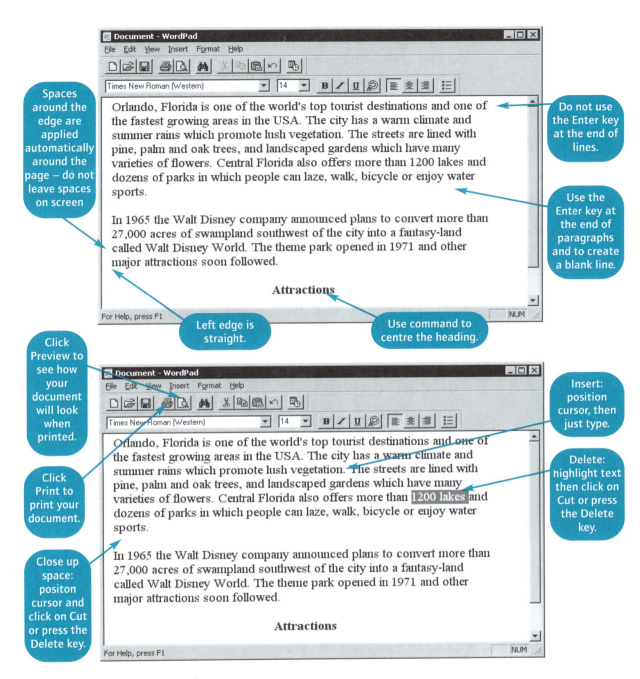

Spaces around the edge are applied automatically around the page – do not leave spaces on screen

Do not use the Enter key at the end of lines.

Use the Enter key at the end of paragraphs and to create a blank line.

Left edge is straight.

Use command to centre the heading.

Click Preview to see how your document will look when printed.

Click Print to print your document.

Close up space: positon cursor and click on Cut or press the Delete key.

Insert: position cursor, then just type.

Delete: highlight text then click on Cut or press the Delete key.

The Learning Curve

Some features of using a word processor

Changing the look of your words

The letters and other characters you see on-screen and on paper can be shown in different sizes and designs. A **typeface** or **font** is a set of characters of a particular design.

Font size

The sizes of characters are measured by an old system called **points**. Books and journals use type sizes between 10 and 12 points; 72 point type means there is about one inch from a character's baseline to the baseline above.

Font design

Font with serifs

Serifs are projections at the ends of letters

The letters that you see on-screen or on paper have a particular look, and each look has a name, e.g. Times Roman or Univers. The different designs can be put into two main families depending upon whether the letters are plain or whether they have little projections called serifs:

- **With serifs:** characters with wide ends, like the words in the paragraph above.
- **Without serifs:** plain characters, like the words in the table on the opposite page.

You can experiment with different looks in different sizes. Professional designers often use two different fonts. If you use too many different fonts, your text will look very busy.

> ## SPECIAL SYMBOLS
>
> There are lots of symbols you can put in your documents. Click on the **Insert** menu, click on **Symbol** and then choose **Symbols** or **Wingdings** or similar. Here are some examples:
>
>

Font style

When you have chosen your font, you can add further variety by having bold and italic versions:

- This font style is regular
- *This font style is italic*
- **This font style is bold**
- ***This font style is bold italic***

Spacing your text

When text is presented for other people to read, its layout needs to have breaks and variety, otherwise it is uninviting to looks at and boring to read. Here are a summary and a few examples.

Layout method	Notes	How to do it*
Margins	Plenty of white paper around the edge of the text usually looks good	Use **File** then **Page setup** and **Margins**
Paragraphs	Paragraphs are groups of sentences separated by a blank line or an indent	Press the **Enter** key at the end of a paragraph
Blank lines	The word processor includes enough space between lines for comfortable reading. But you can add extra lines for special effects	Press the **Enter** key for a new line, twice for a blank line
New page	New subjects or chapters usually start on a new page	Click on **Insert** then **Break**
Text aligned left	Starts each line against the left margin. This is a normal printing layout	Click on **Format** then **Paragraph**
Text aligned left and right (justified)	Starts each line against the left margin and also lines up the right edge of the text. This gives an effect like a newspaper column	Click on **Format** then **Paragraph****
Text centred	Text is arranged in the middle of the page	Click on **Format** then **Paragraph****
Tabs	Tabs are set positions on a line that the cursor jumps to. Useful for lining up columns of figures	Click on **Format** then **Tabs**
Bullet lists	Bullet points are blobs at the beginning of a line. They are useful in lists and to give variety	Click on **Format** then **Bullets and Numbering**
Numbered lists	Numbered lists are like bullet lists, but their items have numbers instead of bullets. The numberes usually count up from 1	Use **Format** then **Bullets and Numbering**
Tables	Full office-based word processors create grids of spaces into which you can place your text	Use the **Table** menu

* This is how to do it with a mouse but there may be keyboard shortcuts, icons and other menus.
** Spaces are added automatically.

The Learning Curve

Advanced features
tables or grids
stored styles or
templates
mail merge
automatic correction
mathematics
password protection

Left aligned

Orlando, Florida is one of the world's top tourist destinations and one of the fastest growing areas in the USA. The city has a warm climate and summer rains which promote lush pine, palm and oak trees, and landscaped gardens which have many varieties of flowers.

Right edge is jagged.

Left edge is straight.

Justified

Orlando, Florida is one of the world's top tourist destinations and one of the fastest growing areas in the USA. The city has a warm climate and summer rains which promote lush pine, palm and oak trees, and landscaped gardens which have many varieties of flowers.

Right edge is straight. This makes extra gaps in the text.

Left edge is straight.

Centred

Orlando attractions include:
Epcot
Gatorland
King Henry's Feast
Pirate's Dinner Adventure
Ripley's Believe It or Not
Seaworld
Terror on Church Street
Universal Studios
Walt Disney World
Wet 'n' Wild

Each line is in the centre of the page.

Bulleted

Orlando attractions include:
- Epcot
- Gatorland
- King Henry's Feast
- Pirate's Dinner Adventure
- Ripley's Believe It or Not
- Seaworld
- Terror on Church Street
- Universal Studios
- Walt Disney World
- Wet 'n' Wild

Each line has a big dot (bullet) in front.

Different document layouts

Checking your work

Computer programs are powerful but they can still be stupid. The spellchecker on your word processor is useful but can only check if the word is in the dictionary. A spellchecker will not notice if you have used the wrong word for the job. Here are some ways you can check your work:

- Ask other people to look at your work
- Wait a while before you check your work
- Print the work and check it on paper

Using graphics

Graphics are any pictures or graphs produced by your computer. Most computers have at least one graphics packages; Windows has Paint.

Paint packages

A **paint package** is a computer program which lets you do freehand drawing and colouring on-screen. You can also take other images, such as a photo, and make changes, perhaps by adding a moustache!

There is a paint program, often called Paint or Paintbrush, included in most collections of computer software; you can use it to achieve the following effects:

- Making simple drawings
- Graphics for reports

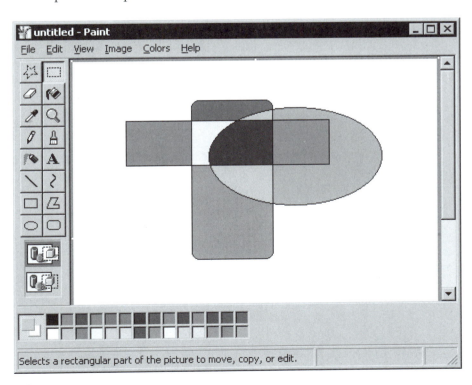

Some graphics
dingbats such as ✔
freehand drawings
clipart
scanned photos
charts and graphs
computer-aided design
3D models
animation

See also: **Using spreadsheets**, page 21.

Paint features
generate lines and shapes
fill with colour
add text
give brush effects
cut, copy and paste
zoom in or out
rotate and stretch
insert pictures

The Learning Curve

- Making changes to photos
- Viewing imported images
- Making computer wallpaper

All paint programs work in the same general way and have the following basic features:

- **Lines and curves** can be set at different thicknesses.
- **Standard shapes** are circles, ovals, squares and rectangles.
- **Pen or brush** can be set to produce lines, blobs, stipples and other textures.
- **Colours** are selected from a palette and poured into shapes.
- **Patterns** can be used instead of colours or with colours.
- **Rubber** lets you rub out mistakes or change your mind.
- **Edit commands** allow work to be changed using undo, cut, paste and copy.
- **Shaping commands** allow you to trim, stretch and rotate.
- **View commands** allow different magnifications.
- **Insert commands** allow pictures and other effects to be brought in from other programs or files.

Communicating

Using email

Electronic mail, or email, is a system on the Internet which allows you to send messages and other computer files to any other user who has a computer connected to the Internet. When your computer is not connected to the Internet, the incoming messages are stored by your internet service provider (ISP).

The ISP has a computer which links you to the Web. You connect your computer to your ISP using a modem and a special phone number. Your computer will have an icon for dial-up connections. Click on this icon to dial into your ISP and check your email. You may need to enter a password. At work, college or school your computers may be permanently connected to the Internet.

An email program is computer software which lets you exchange messages with other users connected to the Internet. You can normally use it for these functions:

- Prepare and store messages before sending
- Send messages to other users
- Send messages to groups of users
- Receive messages from other users
- Scan and read the messages received
- Send a reply to any message
- Print out messages onto paper
- Send and receive computer files with messages
- Delete or store messages
- Store favourite email addresses

Email can be used to send and receive text, pictures, videos and sound. With some software packages it is even possible to make phone calls to another user who has similar equipment. All email programs work in the same general way and have the following basic features:

- **Header** is where you enter the email address of the person to get the message, anyone who is to get a copy, the subject of the message.
- **The body** is where you type the text of the message. You can cut and paste text from elsewhere, such as a word processor.

Online means connected to the Internet.
Offline means not connected to the Internet.

Email programs
Eudora
Outlook
cc:Mail
Hotmail/Webmail

Enter email address here

Attached files will travel with the email

Type your message here

Creating an email with attachments

- **Address book** is where you can store the email addresses of people you know.
- **Signature** automatically adds your chosen signature and other details at the end of the message.
- **Reply options** let you return a message to the sender without typing in the email address.
- **Reply to all** lets you return your message to the sender plus anyone who was on the copy list.
- **Forward** lets you pass the email on to someone new, perhaps adding a comment of your own.

COMPOSE YOUR MESSAGES OFFLINE

If you are paying for your connection time, it is best to write your messages before you go online. You can then take your time and prepare several messages which are stored temporarily in your outbox. When you go online the email software can quickly send mail from your outbox and will get any mail which is waiting for you. New mail will appear in your inbox.

Sending email

- Select the command or icon for a new message.
- Use the **To** field to enter the email address you want to use. This address might be available from the address book icon at the top of the screen.
- Use the **cc** field if you are sending anyone a copy.
- Use the **Subject** field for your choice of short title.
- Use the body area to type your message. You can also use the paste command to import any clip art or other effects.
- To attach a file, use the appropriate command or icon then point to the place in your computer where the file is saved.
- When finished, use the **Send** command. If you are online the email will be sent. If you are offline the email will be sent next time you go online.

Receiving email

- Make sure you are online. For home computing this normally means selecting the icon of your internet service provider (ISP).
- The modem will dial your ISP and you may have to enter your password.
- Once you are online, use the appropriate command to download any messages from your mailbox held at the ISP.
- Any incoming mail is downloaded to your inbox; click on any new messages to open them.
- Read the message and perhaps reply by selecting **Reply**.
- Exit the message then transfer it from your inbox to a storage folder.

Email addresses

There are many millions of people on the Internet and there are many possible email addresses. You need to be careful when reading and writing an address you have been given. Although they may seem confusing, email addresses always follow the same pattern:

- **User name:** this is your personal address; it could also be two names separated by a dot, e.g. jim.jones, or an underscore, e.g. mary_jones.
- **Separator:** found on your keyboard, sometimes called the 'at' or 'axon' symbol.
- **Domain name:** an organisation has to register their particular name.
- **Type of domain:** helps to indicate the type of organisation, such as business, government, network and educational.

Types of domain

.com   commercial
.org   non-commercial
.gov   government
.edu   educational
.net   network
.co.uk UK commercial

Emoticons

:-)   happy
:-(sad
;-)   wink
:-0   shock

Netiquette

Etiquette means good manners, netiquette means good manners on the net. If you are emailing a stranger it is wiser to be polite, just as you would with a letter or phone call. Some suggestions:

- Don't shout with capital letters.
- Always fill the **Subject** field with a helpful title.
- When replying, don't return the whole of a message.
- Check the address before you hit **Send** or **Reply**.
- Wait and think before you reply.

Messages can cause embarrassment if they reach the wrong people or fail to arrive with the right people. You cannot recall emails once they are sent.

COMMON ABBREVIATIONS

AFAIK	as far as I know
BBL	be back later
BTY	by the way
FAQ	frequently asked questions
IMHO	in my humble opinion
RTFM	read the flipping manual
TTFN	ta ta for now

Presenting work

Computers can help you to show your work to other people, such as presenting the results of a project. The word processor, spreadsheet and graphics programs described earlier give you plenty of choices for making your work look good.

Whichever method you use to present your work, it is important to make a good impression. Here are some things you can do to check and improve your work:

- Check that the content is relevant – don't use it just because you have it.
- Select the important information and make sure it stands out.
- Put lots of space between items.
- Beware of using too many fancy effects.
- Use a spellchecker then use a human being.
- Ask other people what they think.

Some presentations

covers for projects
text for projects
project results
announcements
financial information
marketing information
automatic slide shows

Presentation programs

Specialised programs, e.g. PowerPoint, are designed to help you make materials for presentations and then to run the presentations. You can use presentation programs for the following purposes:

LEVEL 2

- Create OHTs or slides
- Store OHTs and slides
- Run a slide show from a computer
- Fade in and out between slides
- Print paper copies of all slides
- Print summaries of slides

Most presentation programs work in the same general way and have the following basic features:

- **Design options** allow you to choose different styles of ready-made slide plus ways of laying out information and graphics.
- **Notes provide** information to go with the slides.
- **Clip art** can be added along with other pictures.
- **Graphs and charts** can be imported from other packages.
- **Edit commands** allow work to be changed using undo, cut, paste and copy.
- **View** generates different types of slide, summaries, speaker notes and handout notes.
- **Slide show** allows you to show the slides full-size on the computer screen or on a projector screen; there is the option to add timing and let the slide show run automatically.
- **Slide effects** include special effects as slides change and added sound effects.

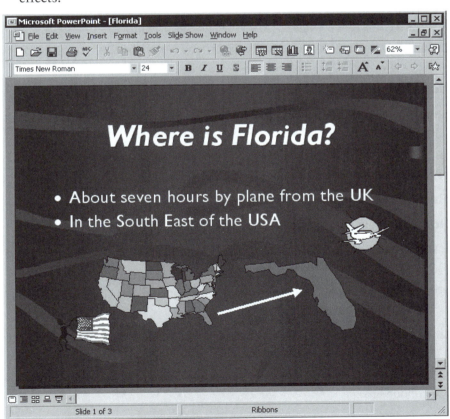

Typical presentation slide combining text and graphics

Glossary

Alignment: arrangment of text compared to margins of paper. Left alignment starts each line hard against the left margin. Right alignment ends each line hard against the right margin. Centred alignment places the text in the middle of the page.

Attachment: a file, such as text, that can be linked to your email message and will be sent with the email.

Bookmark: a shortcut to a favourite website; also called a favourite.

Browser: the software needed to use the Internet. Popular browsers are Microsoft Explorer and Netscape Navigator.

Clip art: drawings and other images you can import into your files.

Cookies: small text files placed on your computer when you visit some websites.

Domain: part of a website address which shows where the site is from, e.g. .com or .co.uk.

Download: copying something from the Internet onto your computer.

Email: or electronic mail that allows you to send information to a particular person or computer connected to the Internet. Messages are stored centrally until you are ready to download them.

FAQs: frequently asked questions.

Flame: an abusive email.

Home: the webpage your browser automatically points to when you log on.

HTML: Hypertext Markup Language is the standard format for creating and showing pages on the Web

ISP: an internet service provider is a company that connects you to the Web and also handles your email.

Link: a word or picture that, when clicked, jumps you to a new place. Also called hyperlinks, they are usually underlined.

Modem: a small box or plug-in card which connects your PC to the Internet over a phone line.

Portal: a website which is designed as a first port of call and which has links to many other sites.

RAM: random access memory is part of the hardware in your computer which temporarily stores information. RAM is measured in megabytes (MB), e.g. 32 MB.

Search engine: a website which finds other webpages when you enter keywords or ideas that interest you. Popular search engines are AltaVista, AskJeeves, Excite, Lycos and Yahoo.

Software: the data which controls computer hardware and allows the computer system to provide services and information.

Server: a high-capacity computer which stores and distributes information such as webpages and email.

URL: URL stands for uniform resource locator but is usually known as the website address.

Part 2: The Bottom Line

This part concentrates on what you must do to get your key skills qualification. It will show you:

- The words and ideas of the key skills.
- The difference between level 1 and level 2.
- How you can practise the skills.
- What must be in your portfolio of evidence.

This part is divided into four sections:

- **What the unit expects:** This section will explain the evidence requirements of the IT key skill, and how to put your portfolio together. Your portfolio is the key to getting your key skill – this part of the book tells you how to choose your evidence and get it ready.
- **Evidence for level 1**
- **Evidence for level 2**
- **Other forms of assessment and evidence:** This section will tell you about the external assessment and how to prepare for it.

Qualifications and Curriculum Authority

The key skills specifications are published by the QCA, and are widely available through schools, colleges, training establishments and awarding bodies. They are also available on the QCA website (www.qca.org.uk).

What the unit expects

What's the difference between level 1 and level 2?

You need to be clear about the level of key skills you are collecting evidence for. This may depend on the GNVQ you are taking. For a Foundation GNVQ the appropriate choice is level 1 information technology. For an Intermediate GNVQ the appropriate choice is level 1 information technology. For GCSE remember that, generally speaking, level 1 is the same as a GCSE at grades D–F, and level 2 is the same as a GCSE at grades A–C. Try to achieve key skills at the highest level you can. However, it's always a good idea to make sure you achieve the key skills qualification appropriate to the other qualifications you are taking.

What is level 1 all about?

The key skills unit at level 1 ask you to show that you can apply your information technology skills and provide evidence for the following two areas:

- Finding and developing information
- Presenting information

You can collect the evidence for each of the two areas from different places if you like. For example, one GNVQ unit, one GCSE or one other qualification might be great for showing that you can find and develop information, but it might not be the best opportunity to show you can present information. A better opportunity might exist for presenting information in another unit or qualification you are doing. At level 1, this doesn't matter and you should use your best opportunities to collect evidence when you can.

How is level 2 different?

You will need to provide different types of evidence for information technology at level 2, the intermediate level. Level 2 is more difficult and will involve showing you can work at a harder level by:

- Searching for and selecting information
- Developing and deriving new information
- Presenting combined information

The first point to note is that some of the evidence requirements are different. You need to show you can work with more difficult sources of information, you have a wider range of information technology abilities and you can produce more complex work.

What's your point?

Regardless of what level you do, your key skills will be about using IT to help you meet your aims or goals. It is not simply about IT in isolation, just you playing around with a computer. It is not enough merely to find and present information using IT. You need to have a purpose for using the information you collect. The IT key skills will keep asking you to use the information for a purpose and you will need to provide evidence that you have a purpose in mind and you can use the IT effectively to meet it. Finding out information for a discussion, wanting to send some important message or information to someone quickly via the Internet, presenting your work in a report for an assignment, all are reasons for using IT. They are all purposes.

Working with numbers and images in IT

The only time you have to show you can work with numbers and images is when you present information. Numbers and images are two examples of information you can search for and develop. But you do not have to search for numbers and images; you could search for text instead.

Numbers

There are various ways to show that you can use numbers in IT. You can present numbers in lists, tables or a spreadsheet; you can use them to create graphs. Creating an invoice and listing laboratory results are two opportunities to present numbers using IT; they would meet the requirements of level 1. Level 2 needs something a little more demanding, like making graphs from spreadsheets.

If you do decide to construct graphs using spreadsheet software or using a special graph package, keep printouts of the numbers you entered. Draft copies are useful ways of showing any changes you made. Write notes on drafts to explain what needs to be changed and make sure that everything is correctly labelled.

Images

Here you have several options to consider. You could incorporate an image into a document that also has text. For example, you could be working on a report that will have a picture or another sort of image in it.

However, the image can be the main focus of your IT activity; you might be scanning an image to work on or you might be sending an image via email.

Differences between level 1 and level 2

The only way that you can be absolutely certain about what is appropriate is to check with your teacher or assessor. Discuss with them what you are thinking of doing and see what they think. Here are a few suggestions that might help you get started:

Numbers

Level 1 requires simple, straightforward work that shows you can present numbers. Straightforward sources of numbers include things like price lists, receipts and timetables.

At level 2 you are expected to have a better grasp of how to use IT to create and present numbers using appropriate formats. At this level you are expected to be able to generate graphs from numerical data or to use spreadsheets to calculate totals and averages.

Images

At level 1 you need to use something like clip art in an effective way, perhaps importing a clip art image into your text to help improve your work. You can use any types of image from any source, as long as you show you can search out and select images then develop and present them to suit your purpose. It doesn't really matter whether you make a report using a graph or diagram, whether you use a photograph, or whether you create or use an image of another kind.

If you are particularly interested in this area of IT, there is nothing to stop you doing other things that might be a little more complicated than level 1 requires. If you can do this harder work successfully it still counts as evidence. Just make sure you have something in your portfolio that shows you can work with images successfully.

There are many software packages that allow you to use and develop images; they range from working with photographs to developing greetings cards. These packages could be used to generate evidence at level 1 or level 2, depending on the activity you do.

At level 2 you need to work with images by using a more advanced understanding of IT. Some of the possibilities may involve specialist software. For example, you might be doing something in your course or job that involves working with IT and images. All the suggestions for level 1 are open to you as well, but you are expected to take them one step further.

Reminder
Look for images and IT you are using in other courses – these can be used to generate evidence. You might be using CAD on an engineering course, or producing coursework including photographs, charts or graphs.

IMPORTANT INFORMATION FOR ALL LEVELS

Whether you are working with numbers or images, the main thing is to show you can incorporate them into your work in an effective way. You don't need to worry about doing really complicated work with numbers or about creating wonderful graphics. You just need to use numbers and images to support your points or to help achieve your purpose. The focus of the key skills is not the numbers or images themselves but how you deal with them using IT. Be simple, be straightforward, be effective – that's the bottom line.

EXAMPLES OF EVIDENCE

Level 1		Level 2	
Number	**Images**	**Number**	**Images**
Using simple grid layouts, e.g. using the tab key, using a table, using a spreadsheet grid. Copying and pasting simple graphs, e.g. bar charts. Using simple formats with numbers, e.g. setting out the result of a calculation, making a simple invoice.	Importing images such as clip art into your files. Creating postcards on email. Exploring software packages that allow you to create and alter images. Saving photographs on computer and pasting them into documents or sending them via email. Creating and using a simple drawing or diagram, either within a word processor or a dedicated drawing program.	Creating more detailed graphs showing more than one thing. Using specialist graph packages. Spreadsheet work that shows you can use formulas to total or work out averages. Spreadsheet work to show you can experiment with different possibilities.	Making changes to drawings and photographs. Creating and sending images on email (postcards, photographs, etc.). Using a specialist drawing package to create an image for your purpose.

The Bottom Line

What about your portfolio?

Building your portfolio of evidence

Your portfolio of evidence is the work you have done to prove to your teacher and others that you can do what the key skills ask you to do. It is the proof you will need to get the key skills award.

A key skills unit is quite a large chunk of work. It is roughly the same size as a GNVQ unit and a little smaller than a GCSE. This means you may have to carry out several different tasks to have sufficient evidence to

Evidence is proof that you can do what is required in order to get the key skills.

show you can meet the key skills requirements. Make sure your portfolio is well organised and make sure the work inside is clear and easily understood. The two usual types of evidence are:

- Signed records by an assessor who saw you using the IT
- Printouts with suitable notes

The assessor will usually be your tutor, and if they are sure that your work is definitely your work, they should be willing to sign you off as having met the requirements. The simplest approach to collecting and keeping your evidence is to use a separate folder or portfolio.

Consider using the following references to organise and label your work:

- Have a contents page that you keep updating as you build up your evidence.
- Keep records of when you collected your evidence and where it came from (e.g. which GCSE or GNVQ unit).
- Get into the habit of writing down the purpose of your work as you collect evidence.
- Use the key skills sections as to divide up your portfolio. At level 1 they are 'finding and developing information' and 'presenting information'.
- Copies of work are acceptable if the actual key skills evidence is part of some work for another course.
- Keep a checklist of all the things you must cover in your portfolio (e.g. in the presenting section you must show you can use one example of text, images and numbers).

Annotate: to add a note of explanation by writing on your documents.

IMPORTANT INFORMATION FOR ALL LEVELS

Collecting and showing evidence of the IT work you did in finding or developing something is often hard to show. The most obvious evidence to have in your portfolio is the finished work itself. However, take time to learn how to use the **Print Screen** function on the computer and keep copies of the different drafts that you do, especially the ones you **annotate**. Both are ways of capturing your development work as evidence in your portfolio.

Don't worry about showing your mistakes; as long as you have detected and corrected them, they too can be evidence. The printouts should be annotated with labels, brief notes and arrows pointing out stages, important features and changes. Presenting original printouts with your handwritten comments is usually the most convincing evidence.

Keeping evidence of your development is especially important when no assessor is available to watch you at work. The aim is to convince the person looking at your portfolio, who didn't see you do the work, that you could actually do all the development you needed to.

Evidence for level 1

While you learn how IT can help you in your work, you should also be able to compare your use of IT with other ways of doing the same work. IT has many advantages but you need to keep judging it and what advantages it has to offer against other possible ways of doing things. You must also work safely at all times and take appropriate care of the equipment. Avoid losing your information and know how to get help when dealing with errors.

Finding and developing information

What you must learn to do

Searching for information from different sources
You will need to have evidence in your portfolio that shows you can use IT to present images or graphical information, text and numbers. Keep in mind that eventually you will have to turn what you collect into text, numbers and images. You will also need to learn how to collect information from IT and non-IT sources.

- **IT sources** could be files or documents stored in computers or saved on disks; you could also find information on CD-ROMs or the Internet. CD-ROM encyclopedias can be a useful source of information. For more information, look at this section in the level 2 evidence.
- **Non-IT sources** can be handwritten notes, catalogues, price lists, diagrams, books, newspapers or magazines.

Choosing relevant information
Once you have collected your information you need to decide what is relevant to you and what is not. This means deciding which parts meet your purpose best. Print out the information you found and read it through, highlighting the key information with a highlighter pen. Give a brief explanation of why you think the information you highlighted is relevant and keep this and your highlighted work as evidence in your portfolio.

See also: **Using text**, page 26.

Entering and developing information

Once you have separated out the relevant information, you need to enter it into your own IT document. You could be copying and pasting text, importing clip art or even adapting information from an original document by saving it under a different name and then cutting out the information you don't need.

Even if you use other documents, you still need to arrange your own work on the page. This is called formatting. There are several computer techniques to help you with formatting. They include tabs, new paragraphs, headings, bold text and centred text. You don't have to do all of them, just the ones that help you present your work clearly. Whichever techniques you choose, use them consistently throughout your work.

You need to work on the information you collect, using your IT skills to format it. Perhaps you have found some interesting statistics or survey results; space them out attractively on the page. You may need to carry out calculations on the computer; the calculator is often found in the **Accessories** menu of Windows.

Spend time developing your information. This means organizing it and making sure it is clear. For example, if your written information (your text) is to be used to find out the answer to a problem, work on it to make sure it does this. Sort the text out into sentences and paragraphs, and give it headings to make it clearer. If you were finding out instructions for making something, get them into a form that you will be able to understand easily. Ask advice to make sure you are clear about the instructions.

Exploring possibilities and alternatives

When you create your information using IT, you need to explore the available possibilities and try out different things. Spreadsheets are a good way to test what happens when you change the values in a calculation. Ask people how best to develop your information and seek advice on the possibilities available to you. There might be many ways to do something, so you need to consider which will be most suitable for your purposes.

When you try an experiment, print the results as evidence. You could even write on your printout to show what you liked or didn't like, or what worked and what didn't work. Always give an explanation of why you think something worked or didn't work. Keep all your printouts and notes as evidence of your development and experimentation.

KEEPING RECORDS

You can keep records for your portfolio by using the **Print Screen** key on your keyboard. This will create a snapshot of your computer screen which you can copy into a new document. Press the **Print Screen** key then open a new document. Go to the **Edit** menu and click on **Paste**. A copy of the screen will appear in your document. For example, if you have successfully used the calculator, or used the drawing toolbar to create an illustration, why not capture what's on screen and keep a copy?

Collecting evidence

What you need to do	Yellowstone Park	Phone numbers
Have a clear purpose for finding and using information. Then get hold of the information you need; make sure you choose the information most relevant to your purpose and enter it into a computer using appropriate formats. Make sure you use any formats consistently (e.g. spaces or tabs). Once in the computer, you must develop the information so it meets your purpose. You have to find and develop information twice, making sure you have a different purpose each time.	To find out some information on Yellowstone Park in the USA and prepare the results in a document that you could eventually turn into a report. You could use a CD-ROM encyclopedia or a Website equivalent. Once you've found the information you need, print it out, read it, select the most relevant information and then type it into a document you've created. Alternatively, you could copy and paste it directly from the CD-ROM. This is an IT source of information. You could then print out the results and use a highlighting pen to mark the information relevant to your purpose. By doing it on paper, you'll have the highlighting plus your annotations and notes to show which information you thought was most important. You could ask some advice about copying images like pictures. Once in your document, you can start making sure the information is clear and well organised. Now work on ways of developing it.	Create a list of useful phone numbers. A spreadsheet is useful for making a list of numbers. You could use the telephone directory and the Yellow Pages. These may also be available on CD-ROM; if you use the paper directories, they are non-IT sources of information. Select the numbers you need. Make a list of numbers you use most or would find useful, e.g. cinema, pizza delivery, doctor. Once you've chosen the numbers you want, enter them into a list and sort them alphabetically. This should give plenty of opportunities to develop the information and explore ways to improve it. You could try different layouts before printing one out to keep by the phone.

The Bottom Line

Evidence requirements in a nutshell

You must show that you can search out and select information using IT then work on it for two different purposes. For each purpose you must find and choose relevant information. You must enter it into the computer using appropriate programs (e.g. a word processor) and use formats to develop and organise it. Explore ways to develop the information to suit your purposes.

HINTS ON FINDING AND DEVELOPING INFORMATION

- Make sure you keep clear and separate records for each purpose.
- Keep a written record of your purpose inside your portfolio of evidence. Explain what you are doing and why you are doing it. This will help people who look at your portfolio.
- Keep clear records of your information sources; make hardcopies if possible.
- Make a list of your chosen formats, functions and techniques, and use it to check you have been consistent throughout your portfolio.

Presenting information

What you must learn to do

Using layouts for presenting different information
You must learn to use layouts that will handle text, images and numbers. Here are a few ideas to get you started.

Examples	Points to consider
Letters	Letters have important rules and conventions. They deal with how to present the address of the person you are sending it to, the date, your address, how to start and finish the letter and the spacing to use. Make sure you know about them
Short reports or essays	Short reports should have clear headings. The text should be organised into correctly formed sentences and paragraphs. The paragraphs should be separated using appropriate spacing
Invoices, memos, newsletters, brochures and other information	All have rules and conventions for presenting text and numbers. You may need to centre your headings
Numbers	Numbers need to be presented clearly with attractive spacing. Provide appropriate column headings and table titles
Images	Pictures may need borders and titles; they may also need to be cropped to fit on your page

Conventions: standard ways of presenting information or laying things out.

There are many ways to present information, so choose the document that best suits your purpose. Before you get started, spend a little time learning about how to lay out the documents you have decided to create. Think about the type of document that will meet your purpose best, then learn how this type of document is normally laid out. This will give you some idea of what is expected when you hand in your work or show it to others.

Consistency in presentations

Be consistent in your presentation; do not jump from style to style or mix up lots of different techniques. Otherwise your work may become confusing and messy. Choose your fonts carefully and take care over tabs, alignments, spacing, lists and any other formatting.

Adapting and developing your work to fit your purpose

This involves taking time to decide what type of presentation or layout will show your work off best. Try to answer these two questions: What am I aiming to do with this information? What do I want to achieve? Then ask yourself, What type of presentation would be best for my purposes? Spend time developing your work so it meets your purposes better.

Make sure you know how to delete and insert information. It also helps if you know how to move text and images and how to organise your work using appropriate headings. Make headings stand out by using bold type or maybe underlining.

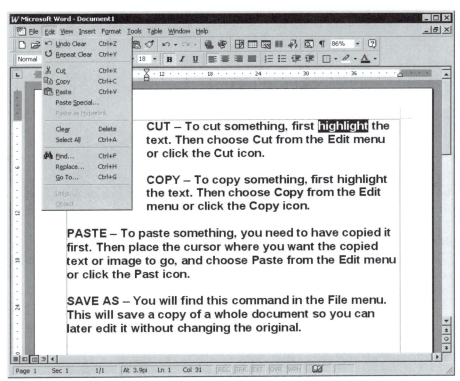

Some basic editing functions of a word processor

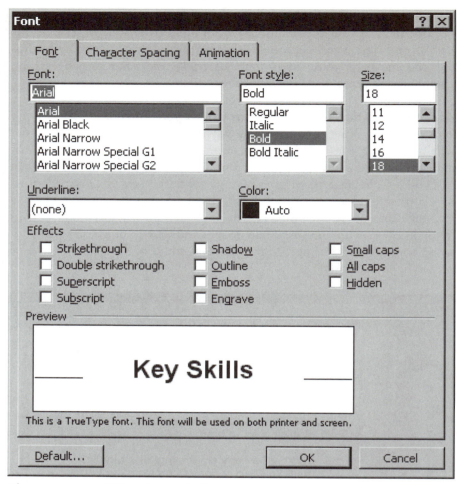

There are many different ways to format your text

Accuracy and being clear

When your information is finally ready for presentation, you need to show you have checked it for mistakes. Most word processors have a spellchecker, but it cannot tell whether you have used a word wrongly.

You also need to show you can use non-IT ways to check your work is accurate and clear. Ask others to check your work then read it through several times. You should double-check that your information is correct. Double-checking is especially important if you have been entering numbers. When you have large amounts of work, highlight small chunks and work on them one at a time.

The importance of saving your work

You need to save your work in an obvious place that will be easy for you and others to find. When you have saved lots of different IT files, make sure you know how to organise them using appropriate file names and by organising them into suitable folders and directories. Ask your tutor about the best way to save your work on a computer. This is because there might be a lot of people using the same computer.

You might have some suggestions of your own. Take notes about how you decided to organise your work (and how you were able to solve the problems of others needing to use the same computer). This will be useful evidence to show how you considered organisation for your portfolio. Make sure directory names and document titles are clear and helpful; base them on the information they contain.

SECURITY

Always have backup copies of your work. The best way to do this is to save your work onto disk. Remember to update your backups every time you make any changes.

Collecting evidence

HOW TO GET YOUR EVIDENCE

What you need to do	Writing a report	Personal interest
Show you can use appropriate layouts to present your information.	You need to write a report containing a picture or lists of numbers. You will have GNVQ/GCSE course work to do that needs to be written up as a report and you could use this as an opportunity to generate IT evidence as well.	Discuss with your teacher or assessor whether you can create evidence in an area that interests you and not related to a course you are doing. Here are some examples:
Spend time developing and organising your work so it meets your purpose.		
Make sure your information is accurate and clear.	This means that your GNVQ/GCSE work will generate the content and you could use your IT skills to present it in a clear and effective way.	• You might have a hobby that has a relevant CD-ROM. It could be anything from a game to garden design.
Save your information in a way that makes it easy for anyone to find.		• You might want join an email chat group and send them a picture of yourself.
Repeat this process using different information for a different purpose.	Work on the organisation of your content. Arrange it under suitable headings, delete any information that you think is irrelevant, and insert new information that helps to improve it.	• You could work with graphics in a particular area.
Make sure your presentation has some text, at least one image and at least one set of numbers.		• You could create greetings cards using clip art or a specialist CD-ROM for making cards.
	Make sure you have an introduction and a conclusion.	Whatever you decide to do, make sure it is likely to meet the actual evidence requirements.
	You could have you headings in bold type.	

EVIDENCE FOR LEVEL 1 | **53**

Carry out spellchecks on your work. Save it under an appropriate name and in a sensible place. Make a backup copy.

You must have a clear purpose in mind. Most of the examples involve presenting, developing, saving and checking information.

Evidence requirements in a nutshell

Remember these two important things when you work on your evidence.

- You need to collect and develop information for two different purposes. The easiest way is to do the work twice but for two different tasks; these tasks could be different units in your GNVQ or different GCSE course work.
- When you do your two different projects, make sure your evidence contains some text, an image and a set of numbers.

In each of your two pieces of evidence you need to show that you can use appropriate layouts to present your information. Present your information clearly and accurately then save it so it is easy to find.

HINTS ON PRESENTING INFORMATION

- Keep the draft copies of your work in your portfolio, especially the drafts that you write on and correct. This is proof of the development work you have been doing.
- Use the **Print Screen** function to print screens that show you have been using templates, layout functions or even the calculator. (Have the calculator showing the answer to the calculation you were doing.) Write on these printouts to explain what you were doing.
- Keep your backup disks with your portfolio; store them in a safe place.

Evidence for level 2

You must learn about the advantages and disadvantages of using IT. This will help you decide when it is appropriate and which technology to use. It will also help you decide which software packages or techniques will be the most appropriate. This will save you wasting time with inappropriate software and routines.

You may also be doing a lot of work on the Internet and using CD-ROMs and other forms of presenting information, so it's important to learn about issues like copyright and confidentiality. As you begin sending files to people via email, you will also need to minimise the risk of viruses.

You must work safely at all times and take appropriate care of the equipment. Avoid losing your information and know how to get help when dealing with errors.

At this level you need to show you can search out and select information then develop and present it. Here are a few more suggestions to help generate evidence for level 2.

See also: **Evidence for Level 1**, page 47–53. It contains lots of useful information and advice.

Searching for and selecting information	Developing information	Presenting information
Using the Find command within a document	Early drafting of documents	Final drafting of documents
Interrogating databases	Pasting into templates	Adding effects and finishing touches
Searching for information on CD-ROMs	Setting up a spreadsheet or database, or creating a word processing file	Working on images, e.g. adding borders or resizing
Searching for information on the Internet	Creating a presentation	Spellchecking
Finding appropriate images to scan	Creating graphs and charts	Grammar checking
Searching for ready-made images like clip art.	Using tables	Working with templates
Using CD-ROM encyclopedias	Cropping and altering images	
Using reference material on CD-ROM	Bringing your information into a computer file you've created	
	Using spreadsheet formulas	

Searching for and selecting information

What you must learn to do

Identifying suitable sources of information

You will need to have evidence in your portfolio that shows you can use IT to present images or graphical information, text and numbers. Keep this in mind when you collect different types of information. You also need to learn how to collect information from IT and non-IT sources.

- **IT sources** could be files or documents stored computers or saved on disks; you could also find information on CD-ROMs or even the Internet. The encyclopedias Encarta, Grolier and Britannica are available online and on CD-ROM.
- **Non-IT sources** can be handwritten notes, catalogues, books, newspapers or magazines. Take time to plan how you will generate the evidence. Consider what types of image, text and numbers you will use and where they will come from. You may decide to scan some pages from these sources and then paste them into your documents.

Looking for information

See also: **Using databases**, page 11.

You need to show that you can look for information using IT. Here are some common opportunities:

- Using the **Find** command in a software package to find a particular person, address or set of words.
- Searching for information on CD-ROM databases, such as the encyclopedias.

See also: **Using the Internet**, page 16.

- Searching for information on the Internet; you will probably use a search engine such as Yahoo. Searching for exact phrases increases your chance of finding what you are looking for, by eliminating documents that include one or other of your search words but not both. Most search engines will search for exact phrases if you put the phrase in quotation marks.

To gather evidence for level 2 you should also be using the operators AND and OR plus the engine's advanced search features.

Selecting information

Once you have collected your information, you need to show that you can select the relevant items. This means looking at the information you collected, deciding what parts meet your purpose best and eliminating what is no use or is not relevant to you. A quick and convenient way of showing this type of evidence is to highlight the relevant information in a printout, explaining why it was important. By doing this you are also showing what is not important (work that is not highlighted).

Collecting evidence

HOW TO GET YOUR EVIDENCE

What you need to do	Environmental tourism	Tourist brochure
Find and choose relevant information for two different purposes Each time you do this, you must show you can: • Decide the information you need to obtain and identify the most appropriate sources to help you find it • Search for the information you need • Identify the information most relevant to your needs from all the information you collect	You want information on where to go to see blue whales. You are doing a project on environmental tourism (ecotourism). The local bookshops and libraries are unlikely to have this sort of specialist information, so you must use the Internet You have selected your search engine (you might even double-check using another search engine). Because you don't want lots of unhelpful information about the colour blue or whales in general, you will use AND and enter 'blue AND whales' in the search box. You could even use the ADJ command to make sure the words appear together and in the right order. You could also use exact phrase searches. Check your web browser to see how this could be done You now need to work through all this information, using titles and content to find out what will be most relevant to you You could print out the best webpages and review their content	To find out about Yellowstone Park in the USA and collect information to allow you to make a brochure. First, you need to get the relevant information into a basic document You could use an electronic encyclopedia on CD-ROM or from the Web Once you have found the information you need, print it out and type the relevant information into a document. You could copy and paste some of the information directly from the CD-ROM Print out the results and use a highlighting pen to mark the information relevant to you. By doing it on paper, you will have the highlighting and your annotations to show which information you thought was most important From the encyclopedia you could copy and paste maps of the USA or Wyoming, where Yellowstone is located Once you have got the relevant information in your document, work on turning it into a form that you can use for the brochure *Continued on page 64*

The Bottom Line

Evidence requirements in a nutshell

You must have evidence that shows you searched out and selected information for two different purposes. As you collect this evidence, you need to show that you can identify the information you need and recognise which sources are likely to be suitable. You must also show that you can carry out searches using these sources. Having found the information you need, you must show you can select the information relevant to your purpose.

Developing information

What you must learn to do

Creating information

This involves entering information and developing it in a coherent way. Before you start, take some time to plan. Think about what your goals are and how you intend to meet them. The main issues to consider are your purposes and who will read your final work. Also think about how these issues will influence the organisation and style of your presentation. It is a good idea to put in your portfolio a few notes that explain your early thinking.

As well as entering information via the keyboard, you should learn how to use techniques like copy and paste, and how to import text, images and numbers from other files to help create the information you need.

Once you have your information in a raw form, you must show that you know how to develop it using appropriate techniques. Developing information applies to your text, image and number work.

Developing text could include creating an appropriate structure for your work, using the page layout, setting margins and organising your work with appropriate headings. At this level you should be using tabs and other spacing techniques to align and position your text, and there are several other techniques you could try.

Developing an image could involve importing an image then presenting it so it best suits your purpose. You could insert the image into a document or a table. You might need to resize it, crop it or customise it. You could be using the **Picture** toolbar to adapt or change your image.

People reading your work will find numerical information much easier to understand as a chart, table or graph. Lists of statistics can be difficult to comprehend and they can interrupt the flow of your text. Here are some examples of creating and developing numbers:

Drawing programs
Microsoft Draw
Paintbrush
Corel Draw
Mac Draw
Mac Paint

- Creating and then customising tables by merging cells or changing the column width. Look in the **Tables** dialog box for the range of possibilities.
- Setting up a spreadsheet. Make sure that the data fields and the data are correct and appropriately organised.
- Doing spreadsheet maths. You will need to select the appropriate formulas.
- Changing the layout of spreadsheets.
- Generating graphs and charts from spreadsheets or special graph packages. You may need to paste them into a document or present them as overhead transparencies (OHTs).
- Creating your own database. There may even be number work here.

Experimenting with your information and exploring possibilities

When you create information using IT, explore the available possibilities and try out different things. When you try an experiment, print the results as evidence. You could even write on your printout to show what you liked or didn't like, or what worked and what didn't work. Always give an explanation of why you think something worked or didn't work.

When you experiment with your work, keep a copy of the originals before you start making any changes. This means you can always return to the original if something goes wrong. If something does go wrong, remember to use the **Undo** function in the **Edit** menu.

Creating new information

This is about using information technology and the information you input to create new information of your own. There are several ways to create new information. You could be representing information in another way for a particular purpose you have in mind. If you are working with numbers you have collected, you could be creating new information by using the information you have found to form your own opinions or to create new statistics, graphs or figures. Here are some examples you could consider.

Be relevant
You do not have to use all the text, image or number techniques.

The Bottom Line

Original information	New information
Number data entered into a spreadsheet	Spreadsheet formulas used to find out the average of the numbers or totals
Number information entered into a spreadsheet	Graphs created to show comparisons using this information and used to help illustrate and support the conclusions you draw
Information taken from different sources	Drawing your own conclusions based on your understanding of this information

Collecting Evidence

What you need to do	Business report	Attendance spreadsheet
Enter and organise your information in a coherent way using appropriate formats to help you Experiment with ways to develop your information to meet your purpose Develop your information and create new information from it, if this is appropriate You need to do this for two different purposes	Create a report on a business; make some recommendations of your own You could create a document; this will eventually become a report for your business course or maybe another course. Open up a word processing package and create a new document. You can make any adjustments by using the page set-up commands and other options There will be a lot of information to enter, including a graph you have created using a spreadsheet package If you would like the reader to see all the relevant information that relates to the graph without having to flick back and forth, then experiment with layout and page set-up. You could consider using landscape pages and a two-column set-up. Try different column widths and font sizes The final section of the report will draw conclusions from the points made in the text and the graph Other things to consider include headings and sub-headings, titles and page numbering	You want to show how many people visited a local attraction You need to create a spreadsheet, customise it and enter the data. You are going to create a graph from this data Add into the spreadsheet the monthly attendance figures from last year. You could use the spreadsheet to give you the average attendance each month You can create a bar chart to show attendance per month and a pie chart to show the relationship between the different groups who attended, e.g. full fare, student/elderly fare and children's fare. You can experiment to adapt the spreadsheet presentation and the graphs. The graphs in particular will have several options to consider

Evidence requirements in a nutshell

Show that you can find ways to explore and develop information and you can create new information. You need to show you can do this for two different purposes, e.g. (1) getting work prepared and ready to be used in a report presented as a document containing a table, an image or a graph and (2) a spreadsheet graph to illustrate the data you have collected.

Show you can use formats that help you develop your information as you enter it and bring it together in a coherent form. Explore ways to develop your information, trying out different ideas to find out what would be most suitable.

HINTS ON FINDING AND DEVELOPING INFORMATION

- Keep a written record explaining the purpose of your portfolio.
- Capture the views on your screen as you develop your work. Use them to record your experiments.
- Keep an explanation of how you created new information in your portfolio. Mention how you created this information and what the information was based on.

Presenting information

What you must learn to do

Using layouts for presenting combined information

You need to show that you can produce work that combines text, numbers and images, not just one type of information. Here are some techniques to consider:

- **Tables:** many word processors will make a table of boxes on the page to help you place your text and numbers; see the **Table** menu.
- **Insert:** word processing programs usually allow you to insert charts, graphs and drawings from other programs; see the **Insert** menu.

You need to consider your reader and your purpose when you make your initial decisions on design, layout and formatting. If you are creating a document to be assessed by a teacher, make sure it is easy to read and has a clear and logical structure. The font needs to be easy to read and its size needs to be appropriate; a 12 point, Times Roman font is a common choice. You could even provide wide margins for teacher comments.

You need to consider how the document should look and begin to lay it out accordingly. There are two ways to approach this: devise your own layout or use a template. A template is a ready-made layout stored in a file. If you are devising your own layout, there are several aspects you need to consider. Software packages normally have a range of templates for different layouts (e.g. letters, memos, invoices), so check what's available to you.

It is important to use a few techniques effectively than to use lots of techniques for the sake of it. You can show the techniques you rejected by keeping screenshots in your portfolio.

You may want to use a pre-existing template instead, or one of the 'wizards' which can help you create good-looking documents automatically.

Page layout features

Page size and orientation (landscape or portrait)

Margins, columns and gutters (space between columns)

Line and paragraph spacing, and tabs

Headings – consider font, size, style

Lists (use numbers, letters or bullets)

Tables and images (clip art)

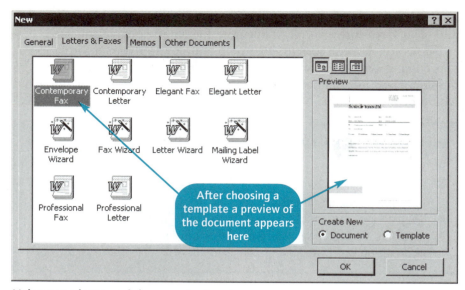

Using templates and document wizards

Making your work fit for purpose

Once you have developed your basic presentation document, you need to show you can refine it. This means adding the final touches to make sure the information is presented in an effective way. There are two things to consider: fitness for your purpose and the effective presentation of your information.

Review and revise the larger issues that relate to your overall goals. You can get a clearer idea of what your document looks like by reviewing it using **Page View** in the **View** menu. Here is a checklist to run through.

- Does it meet my purpose?
- How easy is it to read?
- Can I improve its structure?
- Can I improve its format?
- Are my numbers accurate?
- Are my facts correct?
- Have I been consistent?

You also need to show that you can refine your presentation. You need to consider how to improve its impact and make sure the content is presented in an effective way. There are several things you can consider:

- Fine-tune your formatting and layouts by making sure paragraphs are all properly justified.
- Consider using special features for text, e.g. changing fonts to make headings stand out.
- Look at how the visual impact of images could be increased by text wrapping, borders and shading.
- Use text boxes or call-outs with your images.

Being consistent, accurate and clear in your presentation

Check that you have been **consistent** in your use of key terms and techniques. You can use the **Find** command in the **Edit** menu to see that you have used key words or phrases correctly. The **Find** command can also reveal whether you have used the same words or phrases too many times.

Once you have chosen some layout techniques, you need to use them consistently. Inconsistent layouts can make your presentation confusing and messy. Be consistent over fonts, images, alignments and spacing, paragraph layouts and any other formatting.

Learn how to use the IT checks to their full potential so you can be as **accurate** as possible. Use any spellcheckers (some software may even have a grammar checker). Also, show that you can check your accuracy using non-IT methods.

Copy-editing is best done on paper. When you have a final draft print it out and read it through checking for the smallest of mistakes (missing or incorrect punctuation, or a capital letter that should be lower case). You can then ask other people to proof-read your work, and you can double-check your information. It is especially important to double-check any numbers.

Check your work is **clear**. Don't be too wordy, don't use too much jargon and don't be too vague.

The importance of saving your work

Save your work in a logical place where it can be accessed by those who need to. Consider making your documents easier to find by using the document properties such as title, subject, author name, project name and keywords. You'll find how to do this when you open a new document by

The Bottom Line

HINTS ON PRESENTING INFORMATION

- Keep the draft copies of your work in your portfolio, especially the ones that you write on and correct. This is useful as proof of the development work you have been doing.
- Keep records of all your development work, even the ideas that you didn't use. These show you were experimenting and trying out new ways and techniques.
- Use the **Print Screen** key and the **Paste** command to preserve screens that show you have been using templates, layout functions or the calculator. Write on your printouts to help explain what you were doing.
- Keep a note about why you think your final layout and formats best suit your purpose and explain what you did to make sure your information was presented most effectively. This helps other people to understand your thinking.
- Keep copies of your draft work showing the different techniques you used for correcting your work and tracking your changes.
- Keep your backup disks with your portfolio of evidence.

clicking on **Properties** (in the **File** menu) and then clicking on the **Summary** tab. This is particularly useful if you are sharing directories or folders with other people.

You can also use shortcuts to organise your own files. You could use functions like **My Documents** or Internet **Favourites** to file your work, so it can be accessed more quickly. Find out what organisational short cuts are available in the software you use. When you have saved several different IT files, make sure you know how to organise them using appropriate file names and by using suitable folders and directories.

Ask your tutor about the best way to save your work on a computer. This is because there might be a lot of people using the same computer. You might have some suggestions of your own. Take notes about how you decided to organise your work (and how you were able to solve the problems of others needing to use the same computer). This will be useful evidence to show how you considered organisation for your portfolio. Make sure all directory names and document titles are clear and helpful and based on the information they contain.

Always have backup copies of your work. The best way to do this is to either save your work onto disk or CD-ROM. Remember to update your backups every time you make significant changes to your work. Use headers and footers to identify document and file names, locations where documents are saved, dates of drafts, draft numbers, etc. This can be a useful precaution against loss or mistakes.

Organize your web addresses using the Favourites or Bookmarks functions

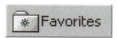

Collecting evidence

HOW TO GET YOUR EVIDENCE		
What you need to do	**Marketing report**	**Tourist brochure** *Continued from page 57*
Present combined information for two different purposes that show you can: • Choose and use appropriate layout techniques to present the information consistently • Adapt and develop your presentation so that it suits your purpose or intentions and the type of information • Make sure your work is accurate, clear and saved in an appropriate way	A report on the marketing success of a product This would be for a business course but all courses will have chances to do essays or reports of some kind You could have sales figures in the form of a graph or table. You could use a spreadsheet to enter monthly sales and average them, work out yearly sales or create a graph. It really depends on the amount and quality of the information you can obtain. This would be an example of combing information	To produce a brochure on Yellowstone Park, Wyoming, USA Use an appropriate template, a brochure or newsletter will do. Explore ways to adapt it Have all the information ready in another document; edit and check the information in this document before you start transferring it to your template

Your combined information must show that you can work with text, images and numbers. You must have at least one example of each

Enter the information yourself. Your report will need to be easy to read and follow, so this will give you plenty of opportunities to consider the organisation, the font and the spacing

Produce paper draft copies to work on and keep your annotations and corrections as evidence of your work

Copy and paste it into the template, review how it looks, make a few changes and edit it again. Print out a final draft and check to see if there are any final changes or corrections to make.

There will be text and images. Find some clip art that shows the states in the USA and a map of Wyoming. A CD-ROM encyclopedia CD-ROM or the Internet will be a good source. You could actually use any sort of graphic that helps get your point across

Crop and work with the images to get them just right. Work on a way to help people spot Wyoming on a map of the USA. You could add arrows by using drawing tools

Evidence requirements in a nutshell

You need to present information for two different purposes. The easiest way to do this is to produce two different IT documents or files for two different reasons. For example, you could produce a report with images or graphs for one GCSE or GNVQ unit and a leaflet, brochure or spreadsheet for a different unit.

You need to have at least one example of using an image, some numbers and text in your portfolio. In each document you produce, you must select and use appropriate layouts for presenting the combined information in a consistent way. You must show that you can develop and refine the work to suit your purpose and to make sure it fits with the type of information you are presenting. Make sure all your work is accurate, clear and appropriately saved.

Other forms of assessment and evidence

External assessment at levels 1 and 2

You will need to take an external assessment as well as produce a portfolio of IT evidence. The external assessment is designed to show that you can work with IT at the correct level under a different set of circumstances. You might be asked to do the assignment in a single long session or in several shorter sessions. This is up to your school, college or assessor to organise for you.

What is the point of an external assessment?

The idea of an assignment is that someone else sets you a series of related information technology tasks, then gives you all the information you need to get on with them. That way you can show that you can carry out different information technology work to complete tasks set by other people. Your portfolio shows that you can use information technology to carry out your own tasks.

It is also attempting to show that you can do larger, related tasks under controlled conditions, such as a time limit, and with someone else setting the tasks. Here is how to look at the portfolio and the external assessments:

- The portfolio shows that you can set and carry out your own information technology tasks to meet your own deadlines and time constraints.
- The external assessments show that you can carry out larger information technology tasks set by other people and meeting the imposed time limits.

When you meet both these requirements, you will get your key skill in Information Technology, and you will have proved that you use IT under different conditions and in different settings.

Part 3: Opportunities

This part highlights opportunities for generating IT evidence in the qualifications you are taking. It will show you:

- How your qualifications can be used to generate IT evidence.
- Where the best opportunities for this evidence arise in the qualifications.

This part is divided into three sections:

- **Evidence from GCSE courses:** You will find this section useful whichever awarding body you are with.
- **Evidence from GNVQ courses:** This section will be useful at both foundation and intermediate level, regardless of whether you are working towards a full award or a Part One award.
- **Information sources:** This section includes a selection of websites which you can use as starting points for research and evidence gathering on the Internet.

The examples provided should be seen as starting points for generating evidence. You will see that some qualifications provide more opportunities than others. However, all contain some opportunities and will at least get you started. Make sure that you take time to read not just your subjects but also subjects that are related to the ones you are taking. This will help you gain a fuller understanding of how and where number evidence can be produced. For example, if you are doing a Business GNVQ then look also at the Business Studies GCSE and the Retail and Distributive Services GNVQ. You may also want to check out the Leisure and Tourism GNVQ.

Vocational awards

The GNVQ Advanced awards are now called Vocational A-levels. From September 2001 GNVQ Foundation and Intermediate awards are likely to be known as Vocational GCSEs.

Evidence from GCSE courses

Art GCSE

About the syllabus

The Art award helps you to develop your creative, imaginative and practical skills as you work in art, craft and design. You will also have the opportunity to explore historical and contemporary sources and make practical and critical judgements and responses.

See also: **Art and Design GNVQ**, page 93.

Topic area 1
Images and artefacts

Searching for and selecting information

The range of the work you produce in art, craft and design will be determined by the variety of media and materials you use. An innovative way in which you can explore and develop ideas is through the appropriate use of information and communications technology (ICT). Images and artefacts can be produced using traditional investigation and research techniques. In addition, IT offers you the opportunity to search and select from a variety of sources, including:

- Files from disks
- CD-ROMs
- Databases
- The Internet
- Scanned material

You may search for and select information on the work of other artists, craftspeople, and designers using appropriate search criteria and operators such as AND but also by using search engines built into Internet packages.

Once located, you will have the opportunity to decide which text, images and numbers are relevant and store them for future use. As you develop, record and store these early ideas, you may wish to establish an electronic sketchbook or visual diary that will act as a resource during your course.

Developing information

You will have the opportunity to use IT for these main purposes:

- As a source of information, exploration and stimulation

- As a means of creative expression
- As an original way of recording, storing, recalling and presenting ideas, images and artefacts

The visual information you store can be brought together in an electronic collage or pastiche. You will need to learn how to use cut and paste, as well as scanning techniques.

IT is particularly useful in developing and exploring ideas quickly because you can easily adjust or amend an image and assess its impact by changing various parameters:

- Scale
- Proportion
- Colour
- Background
- Font and point

Hard copies can be printed to enable you to make judgements about the most appropriate image, and multiple images can be produced.

Many artists, craftspeople and designers use databases or spreadsheets to help them to:

- Store names and addresses for mailing lists (an address book is really a simple database).
- Cost and plan exhibitions or calculate the costs and materials involved in making an image or artefact; spreadsheets are often used to record comments and costs of materials.

Presenting information

There is a difference between using IT for your own purposes as you explore and develop ideas and using it as a means of presenting your final outcome. The first is private and personal and the second is open and public. You must be careful to observe the laws of copyright and confidentiality when you present work that has been developed from a range of electronic sources.

IT offers an exciting way to present your work because it can be transported to a range of locations simultaneously by email and the Internet. Some artists, craftspeople and designers have already abandoned the traditional gallery set-up and established virtual exhibitions of their work using websites.

Once you have decided what you want to present, you can use the software available that best suits your purpose. You could use:

- Static images using drawing programs
- Moving images through animation
- Text-based presentations using PowerPoint.

Make sure that your use of IT is appropriate to the image or artefact you are using. This means that your ideas and creative thinking are helped by the use of the technology and that the technology is used properly and

safely. If you use IT simply because it is available, you may find that it limits your ideas rather than enhances them.

Topic area 2
Personal project

Searching for and selecting information

All art awards expect you to undertake a personal project as part of the course. The programme expects you to develop skills for working across all aspects of art, craft and design, and you will have the opportunity to produce work from direct experience, observation or imagination using any medium, material or technology. IT provides you with an ideal opportunity to extend and enhance your use of materials and processes.

IT provides you with access to a range of sources of information as well as providing the opportunity for you to turn more conventional information, such as drawings, sketches and diagrams, into electronic forms through scanning.

IT can provide the ideal medium for you to:

- Search for and select new visual information
- Sort and store visual information
- Review and record visual information

Developing information

The technology and software available to you will determine the extent to which you can combine and develop visual information. You should have the facility to cut and paste and use other preset formats for layout. You can use the Internet to explore how other artists, craftspeople and designers have chosen to present visual information and use this as a basis for developing your own ideas.

Make sure you save your developments in appropriate files so they can be retrieved later. The development and recording of ideas is an important element in art, craft and design.

Presenting information

You may choose to develop an electronic sketchbook to show the development of your ideas and use it to support your finished project that has been produced in another medium. Alternatively, your project could be entirely electronic. This would enable you to produce:

- An electronic sketchbook to show development
- A database that records the sources of your ideas and information
- A spreadsheet that shows the development time line, materials used and associated costs
- The final outcome itself
- A setting for the final outcome, which could be an electronic portfolio or a virtual gallery on a website
- An evaluation of your work using text

Further opportunities for evidence

Other people's art, craft and design

You are expected to show an appreciation and understanding of the work of other artists, craftspeople and designers and to make connections between their work and yours. These investigations will lead you to look at a range of different electronic sources such as CD-ROMs, gallery websites, electronic books and magazines, and primary sources such as other people's webpages and virtual exhibitions on the Internet. You should look carefully at how creative individuals use electronic media both as a tool to develop their ideas and as a context for exhibiting their work.

Galleries and museums

You will have the opportunity to see examples of art, craft and design by visiting the websites of commercial, private and state-run galleries. Increasingly, galleries and museums are using virtual galleries to promote and advertise their forthcoming exhibitions. An analysis of how images are organised on the screen and how they are contextualised or explained using captions or voice-overs will help you to extend your understanding of the contribution that IT can make to the creative industry.

What you must know
Part 1: The Learning Curve will help you with the knowledge you need.

What you must do
Part 2: The Bottom Line will help you with the evidence you need.

Business Studies GCSE

About the syllabus

The Business Studies award helps you to develop your ability to apply your IT skills to search for, select, organise, combine and use information from a range of different sources. You will be required to evaluate the strengths and limitations of ideas, to distinguish between facts and opinions and to present your conclusions accurately and appropriately.

Topic area 1
The aims and objectives of business

Searching for and selecting information

You will be expected to understand how an organisation works within a range of economic, political and social contexts. This understanding of a business will include the structure, the nature of the business, the aims and objectives and the criteria for judging success. To do this you will need to learn how to select, organise and interpret information from these and other sources:

- Files from disks or other electronic books
- CD-ROMs and other electronic manuals
- Databases and statistical information
- Annual reports or company websites

You are likely to be able to access a wealth of information on individual businesses and companies or types of businesses and companies. You

See also: **Business GNVQ**, page 96; **Retail and Distributive Services GNVQ**, page 121.

Opportunities

Check out the **websites** listed on page 130.

should not simply download and store all information; you must show that you can interpret, decide what is relevant and select for your own purposes. You must make sure you have stored information in an appropriate form. By doing this you will provide yourself with retrievable data for use in evaluating effective business operations.

Developing information

You will have the opportunity to use IT for three main purposes:

- As a source of information using text, images and numbers
- As a means of developing information and combining text, images and numbers for a specific purpose
- As an original way of deriving new information that can be recorded, stored, retrieved and presented

You should ensure that you do not simply cut and paste information from a variety of sources. It is important to provide your own judgements and conclusions on:

- Relationships between business activity and the context
- The development of business structure to meet demand
- The organisation and control of the business
- How the business generates income through market share and how it secures profitability

You should be able to support your judgements and conclusions using charts and tables showing the growth of sales or income. These may be taken from secondary sources such as company reports or produced directly as the result of your own research.

Presenting information

There is a difference between using IT for development and organisational purposes and using it for presentational purposes. Organisation is for you and your ideas; presentation is more public and, as such, your work should follow accepted conventions using appropriate layouts:

- Document formats incorporating margins, columns and headings, and maybe box rules around images and text
- Consistency of text types, fonts and alignments
- Accessibility through clarity, impact, the use of suitable folders and correct file names
- Accuracy of spelling, grammar and punctuation

Make sure that your use of IT is appropriate to your purpose. This means that your ideas, the information being presented or the judgements and conclusions presented should be enhanced by the use of technology, and the technology should be used properly and safely. If you use IT simply because it is available, you may find that it hinders rather than helps your work in business studies.

As you analyse a business and its functions, you should ensure that your analysis is supported by statistical data. The IT key skills qualification requires work to be presented using text, images and numbers. These may be interpreted as using word processing, drawing, databases and spreadsheets.

You should decide whether you wish to present your final report as a hardcopy printout, as an accurately stored and referenced computer folder or directory with a file, or as an electronically delivered presentation using software such as PowerPoint.

Topic area 2
Human resources

The way a business structures itself and the roles, relationships and management are important aspects of your programme. You will be expected to understand and explain the role of human resources development in:

- Recruitment of new staff
- Motivation of the workforce
- Training of staff
- Intra-organisational communication

The types of information you will need to access include:

- Job advertisements in newspapers or websites
- Job descriptions or specifications from personnel departments
- Company intranet sites for details of training courses, programmes or development opportunities
- Memos, notices or other communications in hardcopy formats or electronic formats such as email.

Developing information

You will have the opportunity to develop or use a management chart that shows how an organisation or a team within an organisation is managed. An electronic version of this could link position on the chart with references or links to specific job descriptions, roles within the organisation or advertisements for vacancies.

You will also have the opportunity to explore business or company conventions or regulations for both internal and external communications. These may include:

- Preset formats for memos or letters
- Time taken to reply to internal and external communications
- Response times for electronic correspondence
- Checking or confirmation procedures before an external communication is responded to

Presenting information

An evaluation of a business or company's human resource management provides you with an opportunity to combine text, images and numbers effectively. You may use:

- **Text** to describe the structures and provide examples of different types of effective communication employed for internal and external purposes. This could be an internal memo compared with an external response to a customer complaint.
- **Images** to show management structures and hierarchy using a flow chart or diagram. They could be used to reflect line management responsibility as well as job functions.
- **Numbers** to show annual expenditure on inward investment such as training levels or performance-related pay (PRP) or investment in recruitment through analysis of advertisements for different posts within the organisation.

Make sure the different approaches to presenting information combine effectively within your presentation and make sure that you use the correct structure to balance impact with clarity.

Topic area 3
The management of finance

All effective businesses rely upon the availability of finance. An analysis of a business's or company's annual report will show their performance, including:

- Securing inward investment
- Annual balance sheets
- Future forecasting

What you must know
Part 1: The Learning Curve will help you with the knowledge you need.

What you must do
Part 2: The Bottom Line will help you with the evidence you need.

Although much of this work focuses on the Application of Number key skill, IT provides the ideal medium for combining text, images and numbers to produce an effective report. Used appropriately, spreadsheets can help to make and test predictions, especially where business or commercial circumstances may be changing.

Design and Technology GCSE

About the syllabus
The Design and Technology award helps you to develop your knowledge and understanding of designing so that you can work with suitable design specifications to communicate your ideas. You will be expected to develop your design and making skills by using a range of technologies, by applying knowledge and understanding of relevant processes, materials and techniques, and by using materials, tools and equipment.

See also: **Engineering GNVQ**, page 100; **Manufacturing GNVQ**, page 113.

Topic area 1
Working with a design specification

Searching for and selecting information
The primary source of information is the design specification itself. This may be presented orally, supported by diagrams or drawings or in the

form of a written text supported by other relevant data. It may also on occasions be presented in electronic form in which hypertext is used to link the key areas of the specification. You will be expected to:

- Interpret the specification
- Analyse the requirements
- Calculate the materials
- Identify the technology or equipment
- Prepare a design proposal
- Devise a production plan

Check out the **websites** listed on page 132.

You will need to access a range of sources of information to help you prepare a design proposal or devise a production plan. These electronic sources will include:

- Files for previous examples of work from disks
- CD-ROMs for design ideas
- Databases for lists of available technology including software
- Internet for ideas on suitable materials or production methods
- Scanned material such as diagrams, drawings or production plan templates

Developing information

You should ensure that you have entered, saved and filed all information selected, to enable you to develop your design ideas. You may have access to computer-aided design (CAD) and computer-aided manufacture (CAM) packages. Make sure the electronic information you import is in a format that helps the document.

Use your information to explore the potential of your design proposal. Use modelling techniques to achieve optimum use of materials and record comprehensive information on the final outcome or product including tolerances, materials and making techniques.

Support your design proposal with a flow chart which gives your production plan containing:

- Working schedule
- Decision points
- Timescales
- Deadlines for key milestones for each part of the process
- Alternative methods of manufacture

Use a spreadsheet to identify amounts and sizes and associated costs of materials required. Use it to make and test predictions of costs for one-off, batch or full-scale manufacture of products.

Presenting information

You may wish to provide an electronic version of your design proposal and an electronic model of your final outcome or product. Make sure that such documentation is clearly referenced and accessible using suitable folders, directories and file names.

EVIDENCE FROM GCSE COURSES | **75**

Make sure that your use of technology suits the purpose of your presentation. Use layout and highlighting to add clarity and impact to your design proposal, your production chart and your diagrams or drawings.

Always explore the benefits and disadvantages of using IT. Its effective use at this stage can persuade the client or customer of the strength of your design solution.

Topic area 2
Making skills

Searching for and selecting information

- Before you begin, understand the original specification and assemble all relevant media, materials and equipment.
- Identify suitable sources of information, including hardcopy and electronic versions.
- Search for information using operators and tools.
- Interpret, select and save relevant information.
- Collect all this information and organise it in preparation for developing a production plan.

Developing information

The software available to you will determine the contribution that IT can make to your design and technology work. You should be able to use simple drawing techniques or tables to produce a detailed production plan to underpin the design proposal.

You may also have the opportunity to use a CAD package to help you model your design proposal. This will help you to explore and test out ideas before or during the making process. This process helps you to try out and record alternative solutions and to avoid wasting resources such as time or materials.

Modelling alternative or new solutions enables you to make decisions about effective practice and to support these decisions with printouts or annotated drawings.

Presenting information

You may keep hardcopy printouts of your development work or choose to produce an electronic sketchbook or notebook to show the development of your ideas and use this to support your final outcome or product. It may be that your final outcome or product is also virtual and only exists in electronic form.

Make sure that you have a clear rationale for using IT to support the making process and that the strategies you use are appropriate to the purpose of the project. This will mean that you use and follow established conventions for presentation and that the IT dimension adds to the effective communication of your design and making skills.

Further opportunities for evidence

Class work activities

The learning programme you follow will provide you with the opportunity to explore and practice IT skills. You will have separate tasks that will require you to:

- Search for information in IT-based media.
- Use search engines to locate websites with design information.
- Identify, select and grab relevant information.
- Combine text, images and numbers to support design and production ideas.
- Integrate text, images and numbers to produce and test out new design and production ideas.
- Explore the contribution IT can make to presenting work in design and technology.

Course work assignments

Your programme of study will require you to undertake a project in which you will be expected to combine communication skills, design skills and making skills. This offers you the opportunity to use information technology as a means for effective design communication. IT will enable you to explore different ways to prepare, present and review your design ideas. Your project will provide you with an opportunity to find the most appropriate means of combining your full range of skills to communicate your design ideas effectively and to analyse and evaluate your final product with clarity and accuracy.

What you must know
Part 1: The Learning Curve will help you with the knowledge you need.

What you must do
Part 2: The Bottom Line will help you with the evidence you need.

Geography GCSE

About the syllabus

The Geography award helps you to develop your knowledge and understanding of places, environments and patterns on a local scale and a global scale. You are required to use your developing knowledge and understanding to show an appreciation of the environment and human interaction with it. You will do this by developing and applying a range of skills used in information technology. They include:

- Searching for and selecting information during geographical enquiry
- Developing geographical information in text, images and numbers
- Evaluating, analysing and presenting geographical information on the physical, human and environmental aspects of the subject

See also: **Leisure and Tourism GNVQ**, page 110.

Topic area 1
Geographical enquiry

Searching for and selecting information

You will be expected to develop your geographical skills using these and other sources, both primary and secondary:

Opportunities

- Text-based documents in hardcopy and electronic form
- CD-ROMs containing maps and other geographical sources
- Files on disks
- Databases and spreadsheets containing statistical data
- Webpages on the Internet for current information

Geographical enquiry will often involve work on national, international and global issues. Information technology will enable you to access the most up-to-date information. You must learn to use and apply multiple search criteria and tools such as search engines to make sure you identify information that is relevant to the purpose of your enquiry.

Developing information

You should make sure that you have entered, saved and filed all information selected in a logical and coherent format that helps your enquiry. You should use the medium and associated process, such as copy and paste, to import and combine text, images and numbers. You should then organise this information to help you to develop your ideas. This development may include:

Check out the **websites** listed on page 130.

- Organising your ideas under discrete headings to show the development.
- Introducing tables and charts to explain or support ideas about places and environments.
- Using databases and spreadsheets to present geographical data.
- Using specialist software packages to model or predict outcomes such as climatic changes and impact on sea levels as the result of global warming.
- Creating new information by combining or comparing geographical evidence from different sources.

Presenting information

You will be expected to bring together your descriptions, analyses and interpretations of evidence, and draw conclusions and present findings. Information technology will provide you with a variety of layouts for presenting your evidence.

Always make sure the layout chosen matches the purpose of your presentation. Present your information in a consistent way:

- Use text with suitable fonts and point sizes; do not use too many text types and avoid overusing devices such as bold, italics or underlining.
- Use diagrams and drawings to clarify and support text; make sure they are relevant, sensibly placed and attractively proportioned.
- Statistical information often follows mathematical conventions; it should be clearly labelled and appropriate to the enquiry.
- There is a toolbar to help you check the spelling and grammar in your document; use it wisely.
- Save and reference your files properly to avoid loss of data.

Always explore the benefits and disadvantages of using information technology. Make sure it suits your purposes and gives clarity to your ideas and findings.

Topic area 2
Economic matters

Searching for and selecting information

Economic geography will usually be taught by using a range of secondary sources such as:

- Population figures and distributions, including local, regional and national
- Trade figures, including export and import statistics and exchange rates
- Reports and research papers from national governments and EU directorates

Increasingly, these sources are available electronically on a range of websites. You may be able to trace the website address through promotional or information leaflets, or you may need to use a search engine. School or college computers are usually installed with a search engine. You start by entering some search categories; they can be quite broad to begin with, e.g. European Union. Scanning the addresses it returns and repeating the search process, each time narrowing down your categories, will eventually give you the relevant websites.

When you have downloaded information from the Internet for use, make sure that you have:

- Checked for viruses
- Observed any copyright regulations
- Cleared the data for confidentiality
- Saved the data in a readable form

Developing information

Always try to check the accuracy and currency of your information by cross-checking it with other sources. You are likely to need information from a range of different sources. For example, the economy of a country is often as much to do with its location as with its resources. When exploring the economic interdependencies of places, you will need to combine maps, graphs and diagrams with evaluations or your explanations. Make sure that each piece of evidence is clearly set out and labelled.

Presenting information

You may wish to present your information electronically or print it out as hardcopy. Always proof-read hardcopies before you sign off your work as finished, because errors in the layout may not show up on the screen.

When you create a report from secondary sources you must make sure that your final work is consistent in terms of text forms and layout. This is often best done in hardcopy until you are more confident and familiar with on-screen proof-reading and checking.

What you must know
Part 1: The Learning Curve will help you with the knowledge you need.

What you must do
Part 2: The Bottom Line will help you with the evidence you need.

Opportunities

Make sure the structure of your work is easy to follow. Where it involves a number of different files or attachments, make sure they are clearly labelled and logically set out.

History GCSE

About the syllabus

The History award helps you to develop your understanding of how the past has been represented and interpreted, and how you can use your awareness of this to examine historical resources critically. You are required to use your developing knowledge and understanding of the past, alongside skills of investigation, analytical interpretation and evaluation, to draw conclusions about the periods, societies and situations studied.

Your study of history will help you develop and apply a range of skills that are relevant to the IT key skills. They include:

- Searching for and selecting information during historical enquiry
- Developing historical information in text, images and numbers
- Sorting, editing, reorganising and structuring historical information
- Evaluating, analysing and presenting historical information on events, people and changes in the past

Topic area 1
Chronological understanding

You can study history in a variety of ways, apart from the order in which events happened, but to have a fuller understanding of history you need to be able to place events in their chronological context. You will be expected to develop your historical skills using a range of primary and secondary sources. Primary may be:

- Original documents
- Interviews with people
- Site visits

Secondary may be:

- Other people's records
- Maps, photographs, etc.
- Statistical data

You will have access to many of these sources via the websites of museums and libraries or via CD-ROMs. Make sure you observe the laws of copyright or rules of confidentiality if you download, print or save anything.

Developing information

When you identify relevant information on events, people or historical changes, you should make sure you have entered, saved and filed it in a logical and coherent format that assists your enquiry.

The information you bring together will give you dates of events and times of actions. This order and sequence are very important because they will help you to see how events unfold. They will enable you to sort and develop information into chronological order and sequence, which will help you to understand cause and effect. As you discover who did what and when, you will begin to understand how and why things happened.

Use information technology to help you:

- Sort historical information into chronological order
- Remove irrelevant material
- Add quotations, maps, plans or diagrams to support your ideas
- Insert headings and sub-headings based on dates and events

Presenting information

Your analysis based on chronology will help you to present an explanation of how an event or action took place. For example, you could use information technology facilities and on-screen developments to create a layout that shows how:

- The Spanish Armada was defeated and its ships dispersed around the coast of the British Isles.
- Oliver Cromwell developed from soldier to politician.
- The American states achieved their independence from Britain.
- The sequence of events, or the relationships, that changed the direction of the Second World War, such as the rise of Hitler and/or Stalin.

Your finished work may not always be in the form of an essay or report. You may have the opportunity to make an oral presentation using overhead transparencies or software such as PowerPoint. Make sure the presentation methods you use suit your focus. Chronological order may best be represented by charts, tables or graphs, carefully labelled and annotated.

Whatever form of presentation you choose, make sure that information technology contributes to its accuracy and clarity. Information technology has many advantages over other means of finding and organising information. However, used inappropriately, it can reduce the impact of your ideas.

Topic area 2
Historical sources

Searching for and selecting information

You will be expected to be able to understand and interpret historical events by reading how they were reported at the time. Recent or contemporary history can be easier to access because we have the media and other primary sources to draw on. Generally speaking, the further we go back in time, the scarcer any sources become. They may also be less easy to understand because style, language and spelling have changed over the centuries. Primary sources play a very important role in history, and your

understanding will partly depend on your ability to find information stored electronically.

Through the possibility of scanning original texts and the introduction of museum and newspaper websites, there are an increasing number of sources available to you. They include:

- Texts such as diaries and letters written at the time
- Visual records such as film, video, maps and photographs
- Contemporary documents such as official reports, inventories and research papers
- Contemporary commentaries from newspapers, autobiographies and memoirs

Using the Internet, you can locate sites and search for information by using search engines or promotional or advertising material that provides website addresses. Remember that not all primary sources should be taken at face value because they may be subject to bias or inaccuracy. You should always try to cross-check information with other sources before drawing firm conclusions.

Developing information

When bringing together historical information from a range of sources you will need to sort, edit and reorganise it. This is likely to require more than simply changing grammar or spelling when you seek to develop the information and create a new document from it. You will also need to consider the role of text, images and numbers:

- **Text:** different pieces of writing by different writers produced at different times for different purposes will need to be made consistent by using appropriate icons on your toolbar for spelling, formatting, paragraphing, and standardising typefaces or font sizes.
- **Images:** maps, drawings, diagrams, photographs will need to be selected and labelled carefully and introduced at the appropriate point in reports or essays.
- **Numbers:** databases and spreadsheets could be used to show relevant data and models, or to predict alternative outcomes such as the population size of England had the Black Death not occurred.

Presenting information

Information technology will provide you with access to a wide range of historical sources that will enable you to uncover information about events, people and changes over time. However, your programme expects you to do more than merely describe and explain what you have found out. Your information technology skills should be used to help you select and use a layout that presents your combined information.

You should be able to present information using text, images and numbers to:

- Show your understanding of causes, consequences and changes
- Provide an analysis of key features and characteristics

- Investigate evidence and use it to draw relevant conclusions

Once you are confident in the use of IT, there can be a temptation to use all of the facilities available to you in every piece of work. Remember it is important that others can follow your ideas and understand your conclusions. Do not allow the technology to get in the way of your historical skills by overcomplicating the presentation of your work.

Further opportunities for evidence

Course work

Course work provides you with an opportunity to work on personal projects that contribute to your understanding of history. Chosen carefully, these projects can also provide you with portfolio evidence for your IT key skills at levels 1 and 2. Evidence can be produced by using IT to search for, find and select relevant information and then manipulate and change it into a form that matches your purpose and allows you to present your findings, results and conclusions using a variety of different layouts. Course work is also useful for trying out different types of layout for different purposes in order to discover which is clearest and most helpful in conveying your intentions and clarifying your argument.

What you must know
Part 1: The Learning Curve will help you with the knowledge you need.

What you must do
Part 2: The Bottom Line will help you with the evidence you need.

Home Economics GCSE

CHILD DEVELOPMENT • CONSUMER STUDIES • TEXTILES • FOOD AND NUTRITION

About the syllabus

The Home Economics awards help you to develop the knowledge and skills required to organise and manage resources effectively and safely. You will be able to learn how individuals and groups are affected by the impact of social, cultural and economic factors, as well as by technological developments. You will be expected to apply this knowledge and understanding in order to respond appropriately and effectively to investigations and tasks within the subjects.

The increase in online information and the use of IT in research and development will provide you with an opportunity to develop and rehearse your IT skills. They include:

Check out the **websites** listed on page 133.

- Searching out and selecting information on specific topics
- Communicating and developing ideas with the help of text, images and numbers using specific applications such as email
- Sorting, editing, reorganising and structuring information
- Evaluating, analysing and presenting your ideas, results and conclusions

See also: **Hospitality and Catering GNVQ**, page 106.

Topic area 1
Investigating existing resources

An essential skill for home economics is your ability to plan and carry out investigations and tasks that are set by your programme of study. The skills you use will be the same for different aspects of the subject and you will be expected to gather, record, collate, interpret and evaluate evidence. These skills can be used and developed alongside the different aspects of IT skills:

- Searching out and selecting information
- Developing and presenting information

You will be expected to identify sources of information and evidence and select relevant items. Electronic sources include files on disks, CD-ROMs, databases and the Internet. You will need to use a search engine to find topics on the Internet. For each topic you must choose a suitable starting point, which could include:

- **For child development:** a general survey of webpages, including, those posted by government departments, e.g. Social Services or support agencies such as Childline or the Children's Society.
- **For consumer studies:** a review of current legislation from government webpages, e.g. Customs and Excise, or consumer magazines such as *Which*.
- **For food and nutrition:** a review of current legislation from government webpages, e.g. Environment or Health, or health and fitness agencies such as Weight Watchers.
- **For Textiles:** a survey of webpages, e.g. the Department of Trade and Industry's webpages or ICI's webpages; ICI manufactures commercial textiles and synthetic fabrics.

The exact form of what you retain and store for future use will depend on the nature of your work. If you are exploring, you may wish to store only the website addresses in your address book for future use. If you are downloading more specific text, images or numbers to save for future use, make sure that you are not breaking any copyright laws or confidentiality rules.

Developing information

Make sure that all the information you store has been entered, saved and filed in a logical and coherent format to allow easy retrieval. Use IT and its associated processes, such as copy and paste, to import and combine text, images and numbers. You can then organise your information to help you develop your ideas. Here are some possibilities:

- **Child development:** organise your information and ideas on the development and needs of a child's social, emotional and intellectual abilities, and support them with tables and charts on child growth and development rates.
- **Consumer studies:** investigate online shopping services to compare

information on marketing and advertising techniques used by competing companies.

- **Food and nutrition:** compare how online marketing and advertising of food products are being developed and their impact on traditional methods of influencing individual preference.
- **Textiles:** explore how IT can be used to speed up the design process compared with traditional approaches; in particular, see how it can increase the range and number of different design processes.

Presenting information

Your analysis and investigations will have to be clearly presented to others. This may be via an oral presentation, a written document or an on-screen medium. IT can make a major contribution to the presentation:

- It can help you to explore a range of different layouts quickly and efficiently, enabling you to make a decision about the layout that best suits your information, findings and conclusions.
- It can show the development of your ideas by combining text, images and numbers in order to make your meaning clear.
- It can support different presentation techniques to allow you to present the same information in different ways to different audiences at different times, for example:
  - on-screen format for a small group or to show your tutor
  - electronic information via email and attachments
  - a PowerPoint presentation to an audience
  - traditional text-based hardcopy essays and reports

Topic area 2
Practical projects

Searching for and selecting information

You will be expected to use IT skills to help you undertake your own practical work. The exact contribution these skills will make will depend upon the particular nature of your project.

Your project is likely to be based at least in part on a range of investigations of primary and secondary sources. With the increasing transfer of information into electronic form, IT can be used to identify and collect sources. Here are some possibilities:

- **Child development:** recent scientific research on sensory development and the links between learning and play.
- **Consumer studies:** recent statistics on housing, goods and services, and their links with socio-economic groupings, age and different family groups.
- **Food and nutrition:** recent technological developments and their application in food preparation, processing and storage.
- **Textiles:** recent developments in textile and fabric construction with reference to the latest aesthetic and fashion presentations, shows and styles in Europe and the United States.

Having identified appropriate sources and selected relevant information, you should be able to plan and prepare your project.

Developing information

You are required to produce a series of testable hypotheses to underpin your project. They will need to be explained and supported by the information you have identified and selected. To make effective use of technology, you should:

- Enter and bring together information from your various sources so it is still in a suitable and accessible format. This will require you to change the typeface, font size and paragraphing to help make the material clear.
- Explore the information and make your ideas clear. Check to see how different approaches can change the direction of your ideas.
- Combine text, images and numbers from different sources. Make sure you have been consistent in using any format and labelling conventions.
- Move your ideas forward by adding or creating new information, using spreadsheets to make predictions.

Make sure that what you produce remains consistent with your original hypothesis. Always cross-check the accuracy of information, particularly statistical data.

Presenting information

A practical project is a chance to produce electronically developed information in a form other than printouts of essays or reports. Practical projects that produce an artefact or a final outcome can be supported by a range of different approaches. Here are some examples:

- **Child development:** an educational toy intended to stimulate sensory development could be accompanied by a safety leaflet, or its design specification could be developed using computer-aided design (CAD).
- **Consumer studies:** an electronic guide to good management might use hyperlinks to help users search for and find relevant information on consumer matters. A model website could be established that links to other real website addresses.
- **Food and nutrition:** desktop publishing facilities can produce a menu, a range of recipes or an itinerary for an event or reception.
- **Textiles:** a technical specification for a series of designs could include swatches or mock-ups developed and produced electronically and printed using a colour printer.

The layout you choose, the text quality and your choice of images or numbers, should all help to make your presentation clearer so your audience can follow your lines of reasoning and the conclusions you come to.

What you must know
Part 1: The Learning Curve will help you with the knowledge you need.

What you must do
Part 2: The Bottom Line will help you with the evidence you need.

Mathematics GCSE

About the syllabus

The Mathematics award helps you to develop the ability to use and apply mathematics in solving everyday problems, in thinking clearly and in effective communication. To do this you will be expected to have a good understanding of number and algebra, shape, space and measures, and data handling. You will have the opportunity to develop your mathematics skills through the use of IT by:

- Searching for and selecting mathematical information in software packages and through the Internet
- Developing mathematical information using graphics, geometry and statistics programs
- Analysing and presenting mathematical ideas and information in a variety of ways.

Check out the **websites** listed on page 128.

Topic area 1
Using and applying mathematics

Searching for and selecting information

You will be expected to deal with real problems, simulations and tasks set in a range of contexts. You will need to obtain the most up-to-date and relevant information available. You will have the opportunity to use IT to:

- Identify suitable sources of information
- Search for these sources using operations and tools
- Interpret this information and select relevant evidence

Developing information

Once you have selected relevant information you will be expected to bring it together using the toolbar facilities, such as copy and paste, or other import processes. Always make sure your information is accurately and logically arranged; this will help you to problem-solve or to think mathematically.

IT offers you the opportunity to explore and model solutions using a range of different packages. Here are some examples:

- **Spreadsheets** can test predictions involving numerical or algebraic problems.
- **Graphics programs** can find the most appropriate way to present data, such as histograms, pie charts or line graphs.
- **Geometry programs** can model plane tessellations or problems in 3D space.

Presenting information

IT provides a range of different layouts to help you communicate your ideas accurately, clearly and unambiguously. The packages and programs

Opportunities

available to you will depend on what software has been installed on your computer and the specification of your hardware. Your use of IT should support the development of your mathematical and reasoning skills, not impede it.

Make sure that you always:

- Choose layouts that fit your purpose and support your reasoning.
- Present information in a consistent way. Numbers drawn from different programs, especially fractions, are often presented differently.
- Include in your presentation all the necessary mathematical conventions; not all graphics packages contain scales or labels, so you may need to add them separately.
- Check the accuracy and logic of your outcome; it may be tempting but it is unwise to assume the technology is always correct. Technology is only as good as its instructions.

Topic area 2
Handling data

Searching for and selecting information

Data can be drawn from primary or secondary sources. You will be expected to collect and collate information from a range of sources and combine it electronically. There are two ways to combine data you have collected yourself with data obtained electronically:

- Scan it into the system
- Enter it manually

Either way, you must make sure that any text, images or numbers are properly imported and saved.

Developing information

Information from a variety of different sources will need to be changed into the same format before it can be combined and explored. Use your toolbar, database or spreadsheet programs to create tables or charts to store data.

Data can be in these forms:

- Text which will need to be converted into diagrams or numbers.
- Diagrams or geometrical drawings which may need to be changed in terms of scale, proportion or measurements.
- Graphs which may need to be represented as tables or charts if they are to be combined with other statistics.
- Numbers and measurements which may need to be converted into different formats, e.g. fractions into decimals, centimetres into metres.

Perhaps you are producing mathematical data from existing programs:

- Generating graphs from tables
- Using existing formulas on a spreadsheet

If you need to approximate or estimate the possible outcome in order to cross-check the results. It is easy to make a mistake when you are inputting data, and the quality of the presentation in IT programs is often enough to persuade you that the on-screen information is accurate.

Presenting information

IT will help you to collect, process and represent data in a variety of different forms. You must decide which layout is the one that best suits the data you are presenting. This decision will partly depend on how you want to present your mathematical hypotheses and final interpretation of the data you have analysed.

You will have access to programs that can set out:

- Graphical methods such as line graphs, pie charts, frequency diagrams and histograms
- Tabular methods such as frequency tables
- Charts and diagrams such as networks and probability trees

Your hypotheses and the ideas behind them, together with your supporting solutions and conclusions, should be mathematically accurate and clearly presented using appropriate IT.

Further opportunities for evidence

Class work activities

Your learning opportunities will provide you with a range of individual tasks that can be developed by using IT. The extent to which you use them to provide valid evidence for your IT portfolio will depend upon how you use and record your activities. Your tutor may be prepared to provide you with witness testimony. Make sure you keep this important evidence safe.

Course work activities

Certain programmes of study are based on course work as well as examination. Course work will provide you with an ideal opportunity to work on a personal project. Chosen carefully, it can provide valuable evidence for your IT skills as well as increasing your access to different sources of information and different ways of presenting your results.

What you must know
Part 1: The Learning Curve will help you with the knowledge you need.

What you must do
Part 2: The Bottom Line will help you with the evidence you need.

Opportunities

Science GCSEs: Biology, Chemistry, Physics

About the syllabus

The Science awards help you to develop scientific knowledge and understanding and how to apply them to experiments and investigations. You will have the opportunity to develop an understanding of the power of scientific ideas and the limitations of scientific claims. You will be

expected to appreciate the technological and environmental applications of science within economic, ethical and social contexts and constraints. Throughout your biology, chemistry and physics programmes you will have the opportunity to engage in scientific enquiry which will provide you with a variety of opportunities to develop and use the skills of IT.

See also: **Science GNVQ**, page 123.

Topic area 1
Experimenting and investigating

An essential part of scientific enquiry is your ability to identify a range of primary and secondary sources that can inform and support your experiments and investigations.

Having decided on a specific aspect of scientific enquiry, you will need to investigate sources that can help you to identify relevant experiments as well as appropriate procedures for your experiments. Sources of information will include:

- Text-based documents in both hardcopy and electronic form
- CD-ROMs containing scientific information on materials and their properties
- Files or disks containing records of others' experiments, measurements and observations
- Databases and spreadsheets containing statistical data
- Webpages containing relevant scientific information

Developing information

Here are the things you will be expected to do:

- Identify, enter and bring together relevant scientific information.
- Explore and analyse the information to make sure you can understand and follow a clear line of enquiry or argument.
- Select from others' ideas and observations then develop your own information using text, images and numbers.

Once you have begun to experiment, you will be expected to enter your own measurements, observations and results and save them in an appropriate form. You can derive new information by comparing and combining your measurements with those from different sources. You may also compare your results or conclusions with those of others in order to reach a conclusion and explain your thinking.

Presenting information

Your experiments and investigations will produce outcomes, results or conclusions that will need to be evaluated, explained and, if necessary, used as the basis for predictions. IT can provide you with a range of effective ways to present your findings and predictions as well as support your arguments.

Both your science work and your IT work require you to use a range of different approaches for presentation. Text, images and numbers all play an equal part in explaining and supporting your findings and conclusions.

Make sure your presentation is accurate scientifically and in IT terms, and make sure you use the correct scientific conventions and language in your explanations and your graphical representations.

Topic area 2
Observations, ideas and arguments

Searching for and selecting information

Your programme of study will provide you with the opportunity to explore and communicate the scientific observations, ideas and arguments of others in biology, chemistry and physics. To do this effectively, you will need to obtain information from a range of secondary sources, and read and understand ideas and arguments made using text, graphical materials and numerical data. IT provides one of the most immediate means of accessing the most up-to-date sources of information. You should be able to set up search criteria to locate information from:

Check out the **websites** listed on page 130.

- Technical data in scientific journals and reports
- Text and data in electronic books, CD-ROMs or other files
- Research updates posted on webpages
- Statistical and graphical data from the media and other sources

Developing information

You will have brought together several different types of scientific information from a range of sources. Always check you have been consistent in using scientific language and conventions; this helps to avoid confusion or misunderstanding.

You will need to decide on a common format for this information and then use the toolbar or other processes to standardise your text and separate tables and diagrams so that you can easily follow the ideas and arguments presented when reading on-screen.

If you decide to combine your information to derive new information, make sure the information itself is consistent:

- **Text** should have consistent style and grammar. The pieces you have used may have been written for purposes very different from yours.
- **Graphical information** needs to have well-chosen scales, proportions and measurements. Graphs that need to be compared should be drawn with similar scales, etc.
- **Statistics** should compare like with like. Make sure that categories or ranges are similar in terms of spread or order.

Presenting information

IT offers you a range of layouts for presenting and communicating scientific observations, ideas and arguments. The type of layout you use will depend on what you are attempting to communicate. Here are some examples:

- **A series of facts:** your presentation should be straightforward, simple and easy to follow. You must use the correct scientific language and

follow the appropriate scientific conventions for graphics and drawings.

- **A range of ideas and opinions:** your presentation will need to highlight important points (for example in bold); and according to long-standing conventions, book and periodical titles go in italic, and so do Latin names for animals and plants.
- **An analysis of others' arguments or ideas:** you may wish to use a wide range of graphical and statistical charts and diagrams as you seek to explain and support your judgements.

Further opportunities for evidence

Your learning opportunities will provide you with a range of individual tasks designed to help you explore a range of scientific sources and to help you develop your scientific understanding. You will be able to access and explore the most up-to-date scientific knowledge and information through IT, which may include:

- Using email to question or communicate with others
- Surfing the Internet to explore regularly updated webpages
- Searching for articles in current journals and magazines

You should record your activities and use this record as evidence to support your other work in IT. Take care to observe copyright law and, if applicable, the confidentiality of the information you access. When downloading information from the Internet or a disk, always use a virus checker to minimise the risks of infection.

What you must know
Part 1: The Learning Curve will help you with the knowledge you need.

What you must do
Part 2: The Bottom Line will help you with the evidence you need.

Evidence from GNVQ courses

Art and Design GNVQ

PART ONE • FOUNDATION • INTERMEDIATE

About the specifications

The Art and Design awards include the study of topics such as working with materials, using techniques and technology and exploring other people's work. Remember that all these topics are intended to help you develop your own visual language. Information technology has an important place in your development, both as a source of information and as another tool to help you become more creative.

Topic area 1
Designing and making skills

See also: **Art GCSE**, page 68.

Searching for and selecting information

When designing and making things in art, craft and design, you will need to consider three questions very carefully:

- What am I making?
- Why am I making it?
- Who or what am I making it for?

Your answers to these questions will help you decide on the media, materials, techniques and technology you will need to use. Information technology provides an important source of ideas, information and design processes. Information is available from IT sources such as:

- Files or disks
- CD-ROMs
- The Internet

To use them effectively, you will need to understand how to search for information by file name and by using search engines. The more precise the information you enter, the more likely you'll get the information you require. When designing you will need to look at a range of software packages to find one that is relevant and useful to you.

Opportunities

Developing information

You will have the opportunity to use IT to:

- Prepare a design for a 3D object
- Develop your 2D visual language

IT can be used to:

- Combine different types of images
- Explore and manipulate existing images
- Create new images

IT allows you to create a range of different images and explore a range of different ideas quickly. Make sure you reference, save and store information correctly so you can retrieve it quickly.

Presenting information

IT can be used in several ways:

- To develop your ideas
- To record the development of your ideas
- To present your finished work

IT can be used for all three within the same piece of work, but because each one does something different you must select and use appropriate layouts for each – layouts that suit the purpose and make the meaning clear.

Topic area 2
Exploring other people's work

Searching for and selecting information

You are expected to do three things:

- Compare historical and contemporary art, craft and design work
- Investigate other people's work and express your own views
- Collect, organise and present your findings

IT is an increasing source of information on the visual arts through CD-ROMs on art, craft and design and through Internet sites established by:

- Commercial art galleries
- Public art galleries and museums
- Privately owned collections of visual images
- Virtual galleries created by artists and designers

When searching you need to be clear about the type of information you want and the use it will have. There is so much information available, you need to be sure that what you discover is relevant and useful.

Developing information

You will need to make sure you are not breaking any copyright laws or running up huge costs when accessing information; not all Internet sites are free. When you download information from the Internet, check whether it is compatible. Save it, store it and reference it appropriately.

When you combine information from different sources, particularly text, make sure you do three things:

- Link it appropriately
- Organise it logically
- Format it consistently

Presenting information

Deriving new information by combining a range of existing information, in terms of both text and image, is a very attractive way of producing a substantial essay, report or presentation. Make sure it is fit for purpose:

- Choose the correct layout; a visual presentation and a written report will have different layouts.
- Check for consistency, especially the grammar, when you have combined a range of texts from different sources.
- Use the correct balance between text and images; this will vary between oral and written presentations.

Not all evidence will need to be printed out. You may be able to use:

- Presentation software such as PowerPoint instead of overhead transparencies
- Email with an attachment to send your work to tutors for assessment

Whatever method you choose, select suitable folders and file names to avoid your documents being mislaid.

Further opportunities for evidence

Optional units with art applications

Increasingly, technology is being used by artists to extend their 2D visual language through drawing using a mouse, an electronic pencil and tablet, or a touch screen. Take every opportunity to explore the increasing range of mark-making equipment and techniques offered by technology.

Optional units with craft and design applications

New software packages and hardware with increasing memory have allowed designers to become major users of technology. Packages that allow you to manipulate images and view them through 3D can help you to visualise craft and design ideas without making an object. These processes can be used to present a range of ideas to clients but they are not as time-consuming and there are no materials costs.

What you must know
Part 1: The Learning Curve will help you with the knowledge you need.

What you must do
Part 2: The Bottom Line will help you with the evidence you need.

Opportunities

Business GNVQ

About the specifications

The Business awards provide you with an opportunity to investigate a range of businesses to help you understand how they work, how they develop, and how they are financed.

See also: **Business Studies GCSE**, page 71; **Retail and Distributive Services GNVQ**, page 121.

Topic area 1
Investigating businesses

Searching for and selecting information

Your understanding of business, its functions and aims, will be based on the quality of your research. You will have to identify the sources of relevant information, including:

- Files from disks
- CD-ROMs and other electronic reference sources
- The Internet plus company and research websites

You will need to search for information using operators and search engines. The quality of the information will depend upon the accuracy of your input. Relevant information will include:

- Products and services
- Ownership and management
- Size in terms of number of employees and sites
- Income and expenditure
- Share prices for public companies

Developing information

Your understanding of your chosen business will depend upon the range of information about its business activities that you can access. You will be expected to enter and bring together this information using text, images and numbers.

You will be expected to explore and develop information on:

- Markets and profitability
- Production and distribution costs and retail prices
- Location and distribution
- Investment and finance

As you bring together, explore and develop information, you must make sure it is organised and carefully linked. This offers you the opportunity to combine information from different sources to produce new information such as charts, graphs and spreadsheets, using formulas to calculate costs and profit margins.

Presenting information

Your information can be presented in a variety of forms depending upon your purpose. Here are some aspects you should check:

- Information is displayed using appropriate layouts.
- Information collected from different sources is presented consistently, e.g. the same font size and type are used throughout.
- Different types of information are used to explain and inform your audience.
- Your work is accurate and clear.
- Your work is saved and stored safely to avoid loss of data.

Make sure that your presentation supports the points you want to make. Try not to take the technology for granted. It is easy to believe that the way the work looks is all that matters. Always proof-read and check your work for grammar, spelling and punctuation. Check the accuracy of your graphical and numerical information.

Topic area 2
Enterprise and financial literacy

Searching for and selecting information

You will be expected to show your awareness of the role of finance in business activity by investigating suitable sources of information on:

- The different kinds of accounts available to borrowers
- How to borrow money
- How to get the best deal when borrowing
- The responsibilities of the borrower

IT is now used to provide information on banking services but also online banking. You will be expected to use multiple search criteria to find information from:

- Banks and building societies
- Company reports

Developing information

When you plan and finance a business activity, you will be expected to enter and bring together information on:

- The range of financial documents used by business
- The means of making payments in business
- How to estimate business costs
- How to project profit or loss
- How to buy and sell in business

This will enable you to explore and develop a range of different formats, including:

- Tables or spreadsheets to present numerical information
- Diagrams or charts to present visual information
- Databases or text-based information

Once you have a clear understanding of the financial opportunities available, you will need to compare and contrast the range of services available, including start-up benefits, interest rates and other business benefits. You will be expected to use this information to prepare a business plan based on projected costs and potential sources of financial support.

Presenting information

A business plan provides you with an opportunity to combine text, images and numbers within a single document. Make sure the layout you choose suits your purpose and provides a clear link between the different forms of information.

The final form of your work will depend upon the type of presentation you are making. If you are giving an oral presentation, you will need to prepare overhead transparencies for:

- Text with clear and simple bullet points
- Images such as diagrams or charts that explain processes
- Tables or spreadsheets that examine the financial implications of the business

Always check that you have used the technology appropriately and that you understand how your work has benefited by using IT.

Opportunities from optional units

Sales and customer services

All businesses have customers. An analysis of customer services will include:

- Meeting customer expectations
- Securing customer satisfaction
- Protecting customers' rights

You will have the opportunity to access a range of information using both company and consumer protection websites as well as government information on consumer law and legal requirements.

What you must know
Part 1: The Learning Curve will help you with the knowledge you need.

What you must do
Part 2: The Bottom Line will help you with the evidence you need.

Construction and the Built Environment GNVQ

PART ONE • FOUNDATION • INTERMEDIATE

About the specifications

The Construction and Built Environment awards include topics like the study of towns and cities, how buildings are designed and built, and what materials are used. The optional units give you a choice that includes building design, construction processes, civil engineering, building services and town planning.

Topic area 1
Investigating local areas, buildings and designs

Searching for and selecting information

To investigate the features of your local area and its buildings, you will need to get information, so this is a good activity for using IT. Here are some activities that will produce evidence for the key skill:

- Finding maps, plans, technical drawings available in electronic form on CD-ROMs, databases and the Internet
- Finding text and photos about town features available in electronic form
- Scanning text or images which are not in electronic form

For level 2 you should also show that you have used search criteria such as AND, GREATER THAN, or the equivalent in an Internet search engine.

To make good use of the material you have collected, you will need to explore and develop the information to make it work for you. Here are some possible activities connected to the key skill:

- Choosing the best information for your project
- Importing text and images into your own pages
- Using tabs, centring, etc. to create a layout
- Entering numbers connected with a town, e.g. population figures or traffic figures
- Use IT to make calculations, e.g. adding or averaging in a spreadsheet
- Identifying increases, decreases or other trends

Level 2 expects you to show that you can also:

- Develop your information, e.g. cropping or resizing imported maps
- Organise your material on the page in a consistent way, e.g. with frames or tables
- Think about alternatives, e.g. what happens when the population changes
- Produce graphs or charts from your information

Check out the **websites** listed on pages 130 and 132.

Presenting information

After an investigation you need to show other people your results. Here are some of the opportunities to use IT:

- Keeping your working drafts which show how your work developed
- Printing out maps, drawings and images
- Printing out charts and graphs, e.g. figures on recycling or energy use

Topic area 2
Construction materials, processes and operations

Searching for and selecting information

There are several units which ask you to find out about building materials,

Opportunities

the methods of using them to make buildings, and how people carry out craft operations. You will need information and this will produce evidence for the IT key skill. Here are some possible activities:

- Using technical drawings from CD-ROMs, databases and the Internet
- Choosing materials from electronic or online catalogues
- Looking up figures for materials, such as their strength
- Looking up the costs of materials from electronic or online catalogues
- Looking up the sequence of jobs in project management programs

For level 2 you should also show that you have used search criteria such as AND, GREATER THAN, or the equivalent in an Internet search engine.

To make good use of the material you have collected, you will need to explore and develop the information to make it work for you. Here are some possible activities:

- Important parts of drawings and other images into your own pages
- Using tabs, centring, etc., to create a layout
- Entering the strength of a material or the cost of an operation
- Calculating the number of bricks for a job, etc.
- Calculating the total costs for a job
- Calculating results from your own materials tests

Always check your calculations. For level 2 you could use IT to try out different costs and then choose a suitable one.

Presenting information

After an investigation you need to show other people your results. You should have a good choice of opportunities to illustrate your findings, to show trends and to make comparisons. Here are some typical activities. Be able to explain how using IT made a different to your task.

- Printing out drawings of buildings and schedules of operations
- Showing materials data using bar charts, etc.
- Using graphics to highlight information you think is important
- Checking that your information is accurate and clear

Engineering GNVQ
PART ONE • FOUNDATION • INTERMEDIATE

About the specifications

The topics in the Engineering awards include engineering design and drawing, investigating the workings of modern engineering products and how they are made. Among the optional units are engineering maths and science, computing automation and engineering servicing.

What you must know
Part 1: The Learning Curve will help you with the knowledge you need.

What you must do
Part 2: The Bottom Line will help you with the evidence you need.

Topic areas 1
Producing a design solution
Investigating new technology products

See also: **Design and Technology GCSE**, page 74.

Searching for and selecting information

To investigate a modern engineering product, you will need to get information. Similarly, before you can produce a design solution for someone, you need to understand what is wanted by the client and what is available to use in your design. Here are some activities that will produce evidence for the key skill:

- Consulting product drawings and other technical information from CD-ROMs, databases or the Internet
- Looking up electronic tables and graphs showing the performance of materials and products
- Looking up electronic tables and other details of new technology
- Scanning text or images which are not in electronic form

For level 2 you should also show that you have used search criteria such as AND, GREATER THAN, or the equivalent in an Internet search engine

To make good use of the material you have collected, you will need to explore and develop the information to make it work for you. Here are some possible activities connected to the key skill:

- Choosing the best information for your project, e.g. selecting materials from databases
- Importing text and images into your own pages
- Using tabs, centring, etc., to create a layout
- Entering numbers connected with a material or production process
- Using IT to make calculations, e.g. adding or averaging in a spreadsheet
- Using statistical data, e.g. numbers of people working in an engineering sector
- Trying out alternatives, e.g. a different material

Check out the **websites** listed on page 132.

Level 2 expects you to show that you can also:

- Develop your information, e.g. add detail to a CAD drawing
- Organise your page consistently, e.g. create a table of material properties
- Think about alternatives, e.g. what happens when costs change
- Produce graphs or charts from your information

Presenting information

To show the results of your work to other people, you need to select suitable information and use IT to make it look good. This is especially true when you have produced a particular design and need to persuade other people that it is a good one. Here are some typical activities:

- Keeping your working drafts which show how your work developed
- Printing out CAD drawings and sketches

Opportunities

- Using charts and graphs to compare information, e.g. costs, time and labour for new technology products
- Using graphics to highlight important information from your work

Be able to explain why you think your final results meet the purpose of your work, such as satisfying the design brief.

Topic area 2
Making a product

Searching for and selecting information

There are several units which involve making or servicing engineering products. These activities are planned by using a production plan, following a service schedule and setting up equipment. They provide several opportunities to demonstrate the key skill:

- Consulting drawings, production plans and service schedules on a computer
- Using databases to find material sizes and qualities
- Setting up electronic equipment, e.g. by inputting new scales and settings

To develop this information you might have evidence of the following activities:

- Calculating the measurements from drawings by calculator or computer
- Calculating the quantities and costs of materials on a spreadsheet
- Adjusting equipment settings while removing material

Always check your work and pay attention to layout. For level 2 you should show the use of formulas. Make sure you keep a record of work, such as printouts of your spreadsheets, including any trials.

Presenting information

Keep a record of your work, such as printouts of any spreadsheets. At level 2 you need to produce your own production plan, and this is a good opportunity to use charts and diagrams which show the production processes and their sequence.

Health and Social Care GNVQ
PART ONE • FOUNDATION • INTERMEDIATE

About the specifications

The Health and Social Care awards include these topics: investigating the sector, promoting health and well-being, and personal development and relationships.

What you must know
Part 1: The Learning Curve will help you with the knowledge you need.

What you must do
Part 2: The Bottom Line will help you with the evidence you need.

Topic area 1
Health and well-being

Check out the **websites** listed on page 133.

Searching for and selecting information
This topic is about:

- The needs of others in relation to health and well-being
- Looking at lifestyles
- Considering experiences and feelings
- Identifying health and well-being needs
- Preparing a plan to improve the situation of those under consideration

You will be expected to have a clear understanding of all aspects of good health in general before being able to identify the health and well-being needs of your target individual or group. This will involve you in using electronic sources such as:

- CD-ROMs or other electronic manuals on health
- A variety of websites on health promotion

These electronic sources should be supplemented by other promotional material available from a variety of sources such as clinics, surgeries and chemists. Given that you are expected to consider at-risk groups, you may wish to make contact by email or surface mail with key stakeholders such as the Basic Skills Agency (BSA) or Shelter, who have responsibility for disadvantaged groups.

You will need to use multiple search criteria to ensure that you minimise the amount of information found. The more precise your criteria, the more time you will save in sorting and selecting relevant information.

Developing information
Make sure you enter and save only information that is relevant. This will make it easier to:

- Combine information from different sources
- Set out information in a clear format
- Translate information into different forms

Translating information may mean producing charts or graphs from health data, or creating a database from textual reports. Once sifted and sorted, you can begin to develop a health plan by combining a range of information from different sources:

- Text can identify factors that affect the health and well-being of your target.
- Images can reflect the needs and abilities of your target, such as a child or an adult with learning needs. They may need a range of symbols and signs or a particular layout to make the plan clear and understandable to them.
- Numbers in spreadsheets or graphs can show the benefits of using a health plan, or the dangers of a particular lifestyle.

EVIDENCE FROM GNVQ COURSES | **103**

Presenting information

IT provides you with a range of different approaches for presenting information. You may set out your plan in several ways:

- A report with tables, charts and diagrams
- A diagram or chart using images and notes
- A printed diet-planner with boxes you complete

Whatever approach you choose, make sure it is suitable for its purpose and can be understood and followed by the target person or group.

Try to improve the clarity of your presentation by:

- Proof-reading it
- Checking its spelling and grammar
- Highlighting important sections
- Using sub-headings and appropriate spacing

Topic area 2
Research in health and social care

To carry out a research project, you will need to find and use different sources and types of information. These will include:

- **Primary sources:** information you collect yourself by making observations, asking questions or conducting a survey.
- **Secondary sources:** here are some examples:
  - files, CD-ROMs, the Internet (IT sources)
  - textbooks, promotional material, statistical data (non-IT sources)

You must decide which sources are likely to provide you with the most relevant information and then use the best methods of searching. The clearer you are about what you are looking for, the more precise your search criteria will be.

Developing information

Once you have collected relevant data and information from various sources, you will need to use suitable folders and directories to hold it before sorting and combining. Any research will need to be supported by:

- Background information
- Relevant data
- A clear plan
- A target, focus or purpose

You will need to do three things:

- Explore information in text, images and numbers
- Develop information by combining and organising it to follow your plan
- Derive new information that meets your purpose.

Presenting information

IT offers you a range of different presentation techniques, including:

- Electronic formats to be emailed to your intended audience
- Hardcopy reports in the form of printouts
- Visual presentations of text, images and numbers through PowerPoint

Decide which techniques match your purpose and make sure they are consistent throughout your research.

At first glance electronically produced text can look more powerful and persuasive than handwritten text. However, do not forget to proof-read your work, check the clarity of your images and diagrams or the accuracy of your spreadsheets or tables; any errors will undermine your findings.

Opportunities from optional units

IT is a useful source of information for all aspects of the Health and Social Care programme because it can give you direct access to the Internet. The Internet is the fastest and easiest way to find the most up-to-date information. Webpages are constantly updated by experts working across the world.

With careful use, search engines will direct you to the most recent research. You must take care with this, however, because often the information you find has not been verified by others. Never rely on information from a single source; check it out with other sources.

Take care when downloading information from the Internet; the system contains many viruses. Always use a virus checker before you introduce something new to your computer. Take care to observe the copyright requirements and the confidentiality of your information.

What you must know
Part 1: The Learning Curve will help you with the knowledge you need.

What you must do
Part 2: The Bottom Line will help you with the evidence you need.

Hospitality and Catering GNVQ

PART ONE • FOUNDATION • INTERMEDIATE

About the specifications
The Hospitality and Catering awards include the study of food and drink, accommodation and frontline services, and practical investigations of hospitality and catering outlets and industries. There will be some opportunities for you to specialise in aspects of hospitality and catering that interest you. Most units will provide you with an opportunity to generate some evidence of your IT key skills.

See also: **Home Economics GCSE**, page 83.

Topic area 1
Investigations into industries or outlets

Searching for and selecting information
You will be expected to identify suitable sources of information for your investigation. They can be IT-based sources such as CD-ROMs and the Internet. You can search for information on:

- The size and scope of the industry
- The income and profit generated by the industry

Opportunities

- The size of the labour force and the roles of employees
- Industry trends published in government and CBI reports

The exact nature and amount of information you locate will depend upon the precision of your search criteria and your decisions on what is relevant. Only download and save information that meets your purpose. Always observe any copyright or confidentiality rules.

Developing information

When investigating a local hospitality and catering industry, you will have the opportunity to compare it with other local industries or to compare your local area with the national industry in general. This will give you the opportunity to:

Check out the **websites** listed on page 133.

- Enter and bring together a range of information from different sources, such as data from the websites of national companies and scanned information from local organisations. Try to present this information using a consistent format.
- Explore information by following a line of enquiry using key words or phrases in a search engine to follow up ideas through the Internet.
- Develop information by linking text, images and numbers from a variety of sources to produce a report on your findings.
- Derive new information by using data or statistics to calculate total income for the sector or average earnings for employees.

Presenting information

The size and diversity of the industry will give you access to a great deal of information. The quality of your completed work will depend upon:

- How clear you have been in your purpose
- How careful you have been in selecting relevant information
- How well you have organised your findings
- How successfully you match purpose, findings and presentation

Your investigation into hospitality and catering will give you the opportunity to combine text, images and numbers:

- Text helps you describe and compare your findings accurately and precisely.
- Images help you illustrate your findings; they may be diagrams or promotional material.
- Numbers in spreadsheets or graphs help to support your ideas and findings.

Topic area 2
Investigating accommodation and front office services

Searching for and selecting information

Increasingly, all parts of the industry are using IT to record, monitor and inform businesses of essential information. Here are some examples:

- Residential outlets such as hotels use IT for booking and billing customers.
- Non-residential outlets such as restaurants use IT for monitoring stock such as wines and consumables.
- All outlets use IT for producing staff rostas and wages.

Your investigations should include looking at how different organisations record information and establish systems for its retrieval.

Developing information

Once you have identified or created a suitable programme that will meet your particular needs, you will have to explore how information is entered and combined. This will involve:

- Establishing conventions for consistent input
- Confirming when information should be updated
- Ensuring the model's outputs are accurate and reliable

You may also have the opportunity to observe how:

- Individual pieces of information are combined to provide an overview, as in stock updates, income and expenditure records or projected salary costs.
- Organisations use IT to compare performance overtime or budget for future decisions.

Presenting information

The form you choose for your presentation will depend upon the focus of your investigation. It may include:

- Text that explains your findings through a report or a series of overhead transparencies.
- Numbers in spreadsheets, pie charts or graphs that explain your findings on income and expenditure across different outlets.

Always make sure that your work has been proof-read for accuracy and clarity. Also make sure that IT contributes to your purpose and does not complicate it or confuse your audience.

What you must know
Part 1: The Learning Curve will help you with the knowledge you need.

What you must do
Part 2: The Bottom Line will help you with the evidence you need.

Land and Environment GNVQ

PART ONE • FOUNDATION • INTERMEDIATE

About the specifications

The Land and Environment awards include the study of topics such as caring for animals and plants, investigating environmental factors and the land and environment sector.

Opportunities

See also: **Science GCSE**, page 89; **Science GVNQ**, page 123.

Topic area 1
Investigating environmental factors and looking at ecosystems

When looking at ecology you will learn how environmental factors influence enterprises in the land and environment sector. This topic will provide you with the opportunity to use IT as a means to:

- Identify sources and collect information on your chosen environment
- Record and present data and records on your practical study

Searching for and selecting information

Land and environment enterprises depend upon the local ecology and its relationship to:

- Soil types and characteristics
- Climate (temperature, rainfall, light, wind)

You will need to identify sources of information. These will include:

- Text-based documents in both hardcopy and electronic forms
- CD-ROMs containing environmental and ecological information
- Files or disks containing records of other people's experiments
- Databases and spreadsheets containing statistical data
- Webpages containing weather or environmental data

Developing information

You will be expected to enter, save and combine information relevant to your study:

- Use text to describe the enterprises and provide explanations on the ecology, soil and climate.
- Use maps, plans or diagrams of natural cycles to support your descriptions and explanations.
- Use numbers in charts, tables or spreadsheets to quantify your findings.

As you develop information and use it to derive new data, make sure it is accurate, clear and uses the appropriate format. Your practical study will require you to collect your own data through measurement and observation. This should be organised in clearly labelled charts or tables or in text with relevant headings.

Presenting information

A study can be presented either in written form supported by diagrams and statistical data or in an oral presentation using overhead transparencies (OHTs) or packages such as PowerPoint. Each presentation will require a different approach:

- A written study will need to be clearly set out using paragraphs, headings and appropriate fonts. Keep your layout simple; do not make it too complicated. You can create impact by including boxes around text or by tinting diagrams or graphs. When you introduce images or numbers, make them relevant and explain their purpose to your readers.

- A study that is presented orally relies on two things: what you say and how you support what you say. Set out your OHTs clearly using bullet points; use clear diagrams with good labelling. You may also wish to prepare a brief handout for your audience that covers the main points of your presentation.

Always remember to check your work for errors. It is very important to proof-read the screen or the hardcopy. If you are using software like PowerPoint to present your information, run through your presentation at least once before you do it in front of others.

Topic area 2
Caring for plants or animals

Searching for and selecting information

There are obviously differences in the care of plants and animals at this level but there will be many similarities in your searches for information and the ways you keep your records. Once you have chosen the plant species and propagation methods or the animal species and breeding methods, you will need to identify a range of sources, IT and non-IT, that will help you to:

- Understand genetics in the context of your study
- Plan practical care
- Prepare a health care programme
- Decide on the best way to record data

The National Farmers' Union (NFU) and other organisations have developed a series of commercial software packages to help farmers keep records of their crops and livestock. You may wish to explore them and decide if they are suitable for your purpose or if you can use the model to provide a structure for your records.

Developing information

You will be expected to keep:

- Text-based records that describe and explain developments in your plants or animals.
- Numerical records of growth or yield as well as data on foodstuffs, fertilisers or climatic factors.

As the duration of your caring activity may not follow the full life cycle of your chosen plant or animal, you may wish to use a spreadsheet or other models to project outcomes beyond the life of your project. Make sure that any projections make sense and are in line with the findings of your project.

Presenting information

Your records should meet all the requirements of your project. They should use the correct layout for any information, and they should be

What you must know
Part 1: The Learning Curve will help you with the knowledge you need.

What you must do
Part 2: The Bottom Line will help you with the evidence you need.

relevant, accurate and in line with your findings. Use IT to suit the types of information you have collected. Wherever possible, use visual images and graphics to support your findings.

Where you have collected numerical data, confirm its accuracy before completing your record. If you use a commercial software package, always check that you are not breaking any copyright laws. Where you are using data from a commercial enterprise, make sure you have any permission that is required.

Leisure and Tourism GNVQ
PART ONE • FOUNDATION • INTERMEDIATE

About the specifications
The Leisure and Tourism awards include the study of topics such as investigating leisure and tourism sectors, and marketing and promotion.

See also: **Geography GCSE**, page 77.

Topic area 1
Investigating leisure and tourism sectors

Searching for and selecting information
The leisure and tourism sectors are made up of many different facilities and organisations, from leisure centres and theatres to travel agents and airlines. Before beginning your investigation you should choose a geographical area then list its range of different leisure and tourism outlets. All of them will be suitable sources of information.

- You can find out more about your chosen area by searching for information using a search engine on the Internet. The more precise your search string, the faster the results come back.
- You can find out more about the leisure and tourism components by exploring websites for local government, tourist information offices, English Heritage or the National Trust.

You are likely to identify a great deal of information. You will need to read it carefully and select only data that is relevant to your investigation.

Developing information
Your investigation will need to discover:

- Links between different leisure and tourism operations
- Details of individual leisure and tourism industries
- The activities and facilities provided by each leisure and tourism activity
- The number of people employed in the sectors
- The contribution that leisure and tourism make to the local economy

Some of this information can be collected through a first-hand analysis or survey of your chosen area. More detailed information is likely to be available from your local council's website or published annual report.

IT can provide an ideal vehicle for you to collect, record and combine your findings from a range of different sources. Use all available formats to explore and develop your information. Make sure that:

- Text from different sources is clear and consistent, especially in terms of presentation and the use of technical language.
- Images support your text and that graphics or diagrams are clearly labelled.
- Numbers and statistical data are presented accurately; to make comparisons easier, follow the same conventions throughout.

Presenting information

You may choose to present your findings:

- In a written report
- Through discussion
- Using PowerPoint
- In electronic form

Check that your method of presentation makes your findings clear to your audience. You may wish to use a range of different presentation techniques for the same investigation. This will help you to explore several different approaches.

Do not underestimate the power of graphical information:

- A map can often provide clearer information on an area than detailed text.
- A pie chart of employment in particular sectors often makes more sense than an oral description.
- A database giving brief entries in a series of fields may be easier to understand than a lengthy report.

IT provides you with the opportunity to explore all of these. This does not mean you can forget to:

- Use labels and headings
- Check for consistency and errors
- Check your spelling and grammar
- Use standard headers and footers
- Choose a sensible file name

Topic area 2
Marketing and promotion

Searching for and selecting information

Leisure and tourism organisations will survive only if people know about them and what they have to offer. Traditionally, marketing and promotion have been achieved through:

Check out the **websites** listed on page 130.

Opportunities

- Brochures and leaflets
- Advertising campaigns in newspapers and magazines
- Direct mailing or personal contact

More recently these traditional methods have been supplemented by:

- Videos
- Teletext
- Webpages

You will have the opportunity to explore traditional and electronic methods used by an organisation. This will help you to understand how an organisation works at local, national and international levels.

Developing information

You will be expected to analyse and record different approaches to marketing and promotion. You may wish to:

- Copy and paste examples of web-based promotional materials
- Scan in and save examples of leaflets or magazine pages

Make sure that when downloading from a website you are not breaking any copyright legislation.

You may also wish to explore different costs associated with the various forms of promotional materials, such as:

- The cost of establishing a website including buying a copyright name
- The cost of advertising in different newspapers, magazines or on teletext

Presenting information

Your investigation could provide you with the opportunity to produce your own piece of promotional material based on your findings. You could use IT processes to produce a mock-up of:

- A newspaper advertisement
- A magazine advertisement including colour photographs
- A leaflet to promote an event or facility
- A webpage with an interactive promotion

Make sure you follow established conventions and:

- Keep your message simple
- Provide only relevant information
- Check your words make sense

What you must know
Part 1: The Learning Curve will help you with the knowledge you need.

What you must do
Part 2: The Bottom Line will help you with the evidence you need.

Manufacturing GNVQ
PART ONE • FOUNDATION • INTERMEDIATE

About the specifications
The topics in the Manufacturing awards include investigating how companies use new technology to make products. In other units you

develop your skills for designing and making products. Among the optional units are computing, automation and quality control.

Topic areas 1
Investigating new technology in manufacturing
Working with a design brief

See also: **Design and Technology GNVQ**, page 74.

Finding and developing information

To investigate modern manufacturing you will need to get information about products and companies. Using IT helps make this easy. Similarly, before you produce a solution you need to understand what your design is allowed to use. These activities will help you produce your evidence:

- Using information about companies and their sectors taken from searches of company and government websites
- Looking up website which give information on new technology, e.g. TV tie-ins
- Using databases of product information from manufacturers, e.g. technical specifications
- Using databases and graphs which show the performance of materials and products

For level 2 you should also show that you have used search criteria such as AND, GREATER THAN, or their equivalent in an Internet search engine.

To make good use of the material you have collected, you will need to explore and develop the information to make it work for you. Here are some possible activities connected to the key skill:

- Choosing the best information for your project, e.g. selecting materials from websites and databases
- Importing text and images into your own pages
- Using tabs, centring, etc., to create a layout
- Entering production figures into a spreadsheet
- Using IT to make calculations, e.g. adding or averaging in a spreadsheet
- Using statistical data, e.g. numbers of people working in a manufacturing sector
- Trying out alternatives, e.g. redesigning for a new material

Check out the **websites** listed on page 132.

Level 2 expects you to show that you can also:

- Develop your information, e.g. add detail to a CAD drawing
- Organise your page consistently, e.g. create a table of costs
- Think about alternatives, e.g. what happens when costs change
- Produce graphs or charts from your information

Presenting information

To show the results of your work to other people, you need to select useful information and present it effectively. This is especially true when you

Opportunities

have produced a particular design and need to persuade other people that it is a good one. Here are some typical activities:

- Using CAD or graphics packages to show information by sketches or drawings
- Using CAD or graphics to present a final design solution
- Using computer graphics to help prepare mock-ups, models or prototypes
- Using spreadsheet charts and graphs to compare information, e.g. costs, time and labour for new technology products
- Using graphics to highlight important information from your work

Be able to explain why you think your final results meet the purpose of your work, such as satisfying the design brief.

Topic area 2
Making a product

Finding and developing information
When making a product you will be using a manufacturing schedule and whatever materials and equipment are needed for your chosen product. These activities provide several opportunities to demonstrate the key skill:

- Consulting drawings, production plans and service schedules on a computer
- Using databases to find material sizes and qualities
- Setting up electronic equipment, e.g. by inputting new scales and settings

Presenting information
Keep a record of all your work, such as printouts of trial spreadsheets. For level 2 you need to create a manufacturing schedule, and this is a good opportunity to use tables, charts and diagrams.

What you must know
Part 1: The Learning Curve will help you with the knowledge you need.

What you must do
Part 2: The Bottom Line will help you with the evidence you need.

Media: Communications and Production GNVQ
PART ONE • FOUNDATION • INTERMEDIATE

About the specifications
The Media awards include the development of investigation and production skills.

Topic area 1
Investigating media industries and products

Searching for and selecting information
The media is a large and diverse industry ranging from traditional text-based communications to the latest interactive digital communications

technology. Any investigation of the media will provide a wealth of information from a range of different sources, including:

- Newspapers, magazines and periodicals
- Audio materials, including tapes and discs
- Visual materials such as films, photographs and videos
- Interactive offline materials such as CD-ROMs
- The Internet

The quality and relevance of the information you collect will depend upon your clarity of purpose and your ability to interpret and select information from several sources in various forms.

Developing information

As you collect and record information from different parts of the industry, you will need to:

- Standardise or simplify formats to make exploration easier
- Explore and develop your understanding by trying out different formats

The type of new information you produce will depend upon the purpose of your investigation:

- Audience figures for different media, such as film and theatre, can be compared in a graph or chart.
- Expenditure on an event, publication or film can be set out in a spreadsheet.
- The whole industry can be described using only text, separated into paragraphs, but it will help your presentation to include appropriate tables, charts and diagrams.

Presenting information

As a student on a media programme you will be expected to be able to:

- Select and use layouts that make your meaning and intentions clear.
- Present your findings in a consistent way by making sure that your technical work on layout is appropriate (i.e. paragraphs, fonts and images).
- Explore a range of different approaches that can be used for a range of different audiences.
- Take a professional view of your work by proof-reading and checking it for clarity and accuracy.

Topic area 2
Production skills

Media communications require you to develop an understanding of many different skills, including:

- Video and photography
- Sound production
- Publishing
- Multimedia

You are expected to choose one or more of these areas to investigate then plan and develop these skills to show that you are developing good working practices.

Searching for and selecting information

IT can be used as a primary or secondary source of information, and it can be used as a means of securing first-hand experience on how an interactive medium works. IT as a primary source will require you to:

- Explore the full range of facilities provided by the technology
- Understand that a workstation can operate offline or online
- Become familiar with the limitations of software packages
- Comprehend the opportunities and risks associated with the Web
- Identify the necessary skills for IT-based media and communications

Developing information

Once you have a better understanding of the scope of IT-based communications, you need to build up your skills base. You must learn how to:

- Combine text, images and statistics to greatest effect
- Understand the conventions of layout and presentation
- Choose appropriate text, databases or spreadsheets to suit your purpose
- Communicate effectively using the full potential of the technology

The strength of IT is its potential to explore a range of different approaches very quickly. Use the opportunities it affords to try out a range of techniques and applied skills.

Presenting information

Presentation and effective communication are central to all work in media. This means you must use your presentation to show that you have:

- Understood the importance of presentation
- Worked with your audience in mind
- Checked your results are accurate and considered
- Developed your production skills in your chosen area

What you must know
Part 1: The Learning Curve will help you with the knowledge you need.

What you must do
Part 2: The Bottom Line will help you with the evidence you need.

This topic will provide you with the ideal opportunity to compare the benefits and disadvantages of using IT as a medium for communication. You may wish to contrast IT with film or video and consider this in terms of:

- Access to equipment
- Development time
- Set-up costs
- Audience potential

- Shelf life of products
- Security of information
- Confidentiality of information
- Quality of communication

Performing Arts GNVQ

PART ONE • FOUNDATION • INTERMEDIATE

About the specifications

The Performing Arts awards study topics such as exploring opportunities in performing arts, skills development and performing work.

Topic area 1
Opportunities in performing arts

Searching for and selecting information

You are expected to know and understand the range of opportunities available in the performing arts and entertainment industries. You need to identify the sources that will help you to find out about:

- What the performing arts and entertainment industries are
- What the different job roles are
- The places where professionals work
- Where performances and entertainments take place

A source of general information is the Arts Council's website. It will provide you with a starting point for obtaining information on the range and type of opportunities in the industry.

You will need to analyse the information available, decide what is relevant and use it as a starting point for more specific and detailed enquiry into:

- Theatres, concert halls and cinemas
- Art centres, clubs and community venues
- Travelling venues such as circuses

Developing information

Not all the information you need is likely to be available through IT sources. Commercial organisations, such as cinemas or theatres, or publicly funded venues, such as arts centres, are increasingly using the Internet as a means to promote and market themselves. However, smaller venues or organisations, especially community ventures, are less likely to provide information that is accessible electronically.

Any investigation will need to draw on both types of operation. If IT is to be used as the main form of recording, you will need to:

- Enter and bring together different types of information and reformat them so your final copy is clear and consistent.

- Explore your ideas by following trends or themes across different forms of provision as seen in audience composition, patterns of attendance or viewing figures. Remember to compare like with like when using any form of measurement or statistics.
- Develop your ideas by linking or combining information. This can be done through the use of text, diagrams, tables, databases or spreadsheets. Try out different ways of presenting the same information to see which are the most helpful.
- Produce new information by combining or extending your investigations beyond existing data. For example, take one week and consider the number of individuals attending events in your chosen area; look at the amount of money spent on entertainment; or try to identify gaps in the market and ideas for filling them.

Presenting information

You should have collected a range of different evidence from a variety of different sources on the performance and entertainment industries in your area. This is likely to include:

- Text-based information, including descriptions of the types of industries available and the job roles of professionals.
- Image-based information, including maps of the area identifying the location of industries and plans of different venues.
- Number-based information, including audience, employment and financial statistics presented in tables, charts and graphs.

IT can help you organise and present your findings clearly, simply and effectively. To get the best from IT, you must match your presentation to your purpose and your audience. You can present your material as:

- A printed report
- Using handouts and OHTs
- On disk or in a file
- Through PowerPoint

Whichever method you choose, make sure your work is sound. Proof-read and check your work for consistency and accuracy; ensure it is appropriately saved and readily accessible.

Topic area 2
Promoting, organising and evaluating events

Searching for and selecting information
Use information technology to:

- Plan, prepare and promote the event
- Manage and run the event
- Evaluate the event

To plan, search for information on:

- Costs
- Laws and licences
- Effective advertising
- Ticketing
- Budgets

To manage, search for information on:

- Customer care
- Health and safety
- Equipment hire

To evaluate, search for information on:

- Encouraging audience feedback
- Collecting information
- Presenting your findings

This is a team activity and the team will need to make sure that everyone is clear about the team's purpose and the role of IT, and each team member will need to understand their individual contribution.

Developing information

IT can be used to support:

- **Planning:** use a spreadsheet that identifies the range of activities, associated expenditure, potential income and potential audience size.
- **Promotion:** design posters, leaflets and tickets.
- **Management:** develop information sheets on job roles, customer care and any emergency actions.
- **Evaluation:** use questionnaires or other audience response leaflets as part of the consultation exercise.

Make sure the information you produce is consistent by:

- Using a standard title across all material
- Using the same style for all documents

Presenting information

The information you produce when organising an event has two main audiences:

- The event organisers
- The people attending the event

The main content of this information may be the same but you must make sure that each audience knows what is expected of them. Here are two examples:

- Promotional material must inform people where and when the event is to take place and how much it will cost. The expectations of the booking office and the person buying the ticket are different but both must get the correct date, time and price.

What you must know
Part 1: The Learning Curve will help you with the knowledge you need.

What you must do
Part 2: The Bottom Line will help you with the evidence you need.

- Evaluation responses are more important to the event organisers than to the people who attend. The questionnaire needs to be carefully worded and distributed, so as not to cause offence.

The success of your event will often depend upon the quality of the information you produce. If used properly, IT provides an ideal means for developing and presenting high-quality products.

Make sure that you do not rely on the technology to remove all errors. Always check for consistency of presentation, clarity of information and accuracy of spelling and grammar.

Depending upon the software available, you may be able to use IT to collate your evaluation findings and present them graphically when you record your contribution to the event.

Retail and Distributive Services GNVQ

PART ONE • FOUNDATION • INTERMEDIATE

About the specifications
The Retail and Distributive Services awards investigate the retail and distributive services sector, merchandising and display, and sales and finance.

See also: **Business Studies GCSE**, page 71; **Business GNVQ**, page 96.

Topic area 1
Investigating the retail and distributive services

Searching for and selecting information
Retail and distributive services represent a large part of our economy, and in the UK more people work in retailing and distribution than in any other sector. A very diverse sector, it includes:

- Every type of retail outlet from the corner shop to the supermarket
- Different approaches to shopping and buying
- The distribution chain and main methods of distribution
- The classification of different types of goods
- The legislation that governs the sector

You will find no difficulty in identifying sources of information. Your biggest challenge will be to sift, sort and identify relevant information. The search criteria you use will depend upon your purpose. If you use the shopping basket approach, you can search:

- Websites of retail outlets where goods are purchased
- Websites of the companies whose products you have purchased
- Websites of the distributive service responsible for distribution

Although organisations are increasingly setting up webpages, you may find that some information is not yet available electronically. You will then need to collect it from:

- Company reports
- Observation or interviews
- Catalogues or brochures

Information can be combined by downloading items from the Internet and by scanning paper documents into your computer.

Developing information

Your study will involve you in collecting a great deal of information that will need to be:

- Combined and streamlined
- Made consistent in terms of structure and format
- Organised to follow the development of your study
- Structured in terms of text, images and numbers.
- Presented to suit your audience

Structuring items may require you to turn some text-based information into a chart or table, or create a database for each of the products you have chosen. A map that shows the distribution chain for your goods can be supported by a table that gives the distribution costs at each stage. You may be able to develop a formula to calculate or show the initial and final cost of a product from the beginning of the chain to the retail shelf.

Presenting information

The challenge you face is to present a range of information so that your interpretation and meaning are clear. IT gives you several ways to bring consistency to your presentation:

- Use layouts to support combined information, such as headings and sub-headings, borders around maps and diagrams, highlighted or tinted tables and detailed spreadsheets.
- Standardise text from different sources in terms of style, font and size.
- Have the opportunity to review and revise your work during drafting in order to increase its clarity and impact.
- Use on-screen and hardcopy versions to check for clarity and accuracy.

You may also be aware of some disadvantages with IT and you should write them down to show that you have evaluated your IT use.

Topic 2
Buying, selling, sales and finance

Searching for and selecting information

The main objective of any business is to generate a profit. Retail and distributive services are no different from any other business. Your investigation will require you to gather information on how an organisation:

- Buys raw materials or products
- Promotes its products and services
- Operates at the point of sale
- Finances its activity

Information for these areas will usually be found in the annual reports made by companies. You may have to work with hardcopy versions, although organisations are increasingly publishing their reports on the Internet.

If you are investigating a high street retail outlet, IT can be used to process and compile your findings collected through enquiry, observation and discussion. A more appropriate use of technology may be to explore home shopping or online shopping. A list of shopping sites can be found by looking at newspaper or magazine articles or advertisements giving their email addresses and websites. If you choose to look at online shopping, make sure you have permission from the companies you study.

Developing information

You will need to scan in copies of any relevant documentation, such as:

- Contract letters
- Invoices
- Bills of sale
- Purchase orders
- Monthly financial reports
- Annual profit and loss accounts

Make sure you understand what each document is for. Familiarise yourself with how each company does its buying, selling, paying and recording. You could provide a small amount of text to confirm your understanding, especially the difference between income and profit.

Where a company stores financial data on computer, see if you can use it in a spreadsheet to predict annual or monthly returns. Use IT to turn your raw data into graphs or charts. Discuss your graphs with the company and find out whether they make sense in terms of its business needs.

Presenting information

Look at how online retailing provides a clear pathway to the potential buyer. See how clearly the way to buying is set out and how easy it is to order products. See if you can introduce some of the ideas from online shopping into your investigation.

Use your experience of sales administration and computer-based systems to investigate virtual shopping on the Internet. Outline its strengths and weaknesses and use them to inform your evaluation of IT within the retail and distributive services sector.

Science GNVQ

PART ONE • FOUNDATION • INTERMEDIATE

About the specifications
The GNVQ Science awards study topics such as applying practical skills in science, experimenting and carrying out scientific work, and applying scientific knowledge, skills and understanding.

Topic area 1
Measuring, observing and applying practical skills in science

See also: **Science GCSE**, page 89.

Searching for and selecting information
To obtain useful findings and meaningful conclusions, scientific work requires:

- Thorough planning
- A clear purpose
- Accurate observations
- Careful recording of results

IT offers you the opportunity to explore a range of scientific sources that will provide you with the most up-to-date information. It will also provide you with the means to record and organise your observations and measurements in a flexible format to help you develop your ideas and reach conclusions.

Check out the **websites** listed on page 128.

Developing information
You will need to decide how you will bring together your observations or measurements and what layout most suits your purpose. This may include:

- Notes of your methods and observations. Keep them clear and simple. Use headings and sub-headings; highlight important sections.
- Drawings and sketches of equipment and observations. These may be entered on-screen or made on paper then scanned in.
- Numerical data or measurements made during experiments. Straightforward records may be recorded in simple tables or charts.
- Experiments to generate data, to derive new information or to support projections may be set out in a spreadsheet using a formula, or they may be used to develop line graphs.

Presenting information
The way in which you set out your results and the methods you use to support your conclusions will help you to meet your purpose. IT offers you a

Opportunities

variety of ways to present information. Use it wisely so that it makes an effective contribution. Here are some of the things IT can do:

- Bring text-based documents to life by allowing you to highlight important sections, varying text type and fonts
- Introduce drawings, sketches and charts to explain processes and outcomes
- Present numerical data in a variety of forms, e.g. spreadsheets, graphs or tables, to increase the impact of your results

Scientific measurements, observations and recordings have to reach a high standard of accuracy. Make sure that you bring a similar standard to your work on IT by proof-reading, by checking for clarity and by following the usual conventions for labelling graphs and tables.

Topic area 2
Scientific knowledge, understanding and skills

Searching for and selecting information

Your GNVQ will provide you with a range of opportunities to explore developments in the world of science. There are several ways of using IT to explore your chosen topic:

- Researching its history by using CD-ROMs
- Investigating recent developments on commercial websites
- Discussing data by exchanging emails

IT gives you immediate access to a range of information. Your searches will need to be as precise as possible. Try to use strings to locate the most relevant information and make sure you are not breaking any copyright laws before you download information from identified sources.

Developing information

The development of information will depend upon:

- The types of sources you have used
- The types of information you need
- The types of information you have collected
- The purpose of your investigation
- The audience for your findings

Scientific information is normally presented in a variety of forms:

- Text-based descriptions and records of methods, chemical reactions, observations and conclusions
- Drawings, sketches or photographs of apparatus and observations
- Numerical data used to quantify reactions or changes

You may need to change the format or layout of the information you collect if you are combining it. This can involve:

- Using headings or sub-headings to introduce consistency
- Turning industrial processes into laboratory experiments and drawing the apparatus

- Converting numerical data into graphs, or vice versa, to compare findings from different sources

At each stage, check that your changes or interpretations are accurate, especially when using formulas or when converting from one type of unit to another, e.g. Fahrenheit to Celsius.

Presenting information

When you explain the principles and workings of your chosen scientific processes or your understanding of other people's research, you will need to combine and present your findings consistently. Remember to choose carefully from the full range of IT processes available to you:

- Use text to make meanings clear
- Use images to make processes clear
- Use numbers to make outcomes clear

Using IT does not mean having to use *all* the techniques available. Try to keep things simple:

- Avoid text with a huge variety of fonts or highlighted sections
- Avoid diagrams with a confusing amount of detail
- Avoid spreadsheets or graphs that are difficult to interpret

You may wish to compare the benefits of using IT in a project with more traditional experimenting and recording processes. You could make comparisons using evaluation criteria such as:

- Ease of access
- Quality of information
- Range of processes available
- Contribution to understanding
- Impact on your audience

What you must know
Part 1: The Learning Curve will help you with the knowledge you need.

What you must do
Part 2: The Bottom Line will help you with the evidence you need.

Opportunities

Information sources

The following Internet websites can be useful starting points when searching for information. They are grouped in topic areas but there are many overlaps. Internet addresses can become outdated and you should aim to keep a list of favourites that are useful to you.

   If you can't find the information you are looking for, you should use one of the search engines described in Part 1.

See also: **Using the internet** page 16.

General

www.ngfl.gov.uk
National Grid for Learning
A useful portal to many sites education and learning sites in the UK

www.gsce.com
GCSE Answers
Offers practical help in GCSE English and mathematics, plus links to other websites.

www.bbc.co.uk/education
BBC Education
Includes useful information on many subjects and is a portal to a range of other websites.

babelfish.altavista.com
Translation website
Enter (or paste) your text and the website will translate to and from major languages.

www.learnfree.co.uk
Learnfree
A website for teachers, parents and learners, including useful tips about qualifications and studying.

English

www.ngfl.gov.uk
National Grid for Learning
A useful portal to many education and learning websites in the UK.

www.gsce.com

GCSE Answers

Offers practical help in GCSE English and mathematics, plus links to other websites.

www.plumbdesign.com/thesaurus

Visual Thesaurus

An interactive, animated thesaurus which displays related words on-screen and allows you to navigate connections between them.

daphne.palomar.edu/shakespeare

Shakespeare and the Internet

Timelines, biographical details and links to many other Shakespeare-related websites.

Mathematics

www.ngfl.gov.uk

National Grid for Learning

Useful portal to many education and learning websites in the UK.

www.gsce.com

GCSE Answers

Practical help in GCSE English and mathematics, plus links to other websites.

www.anglia.co.uk/education/mathsnet

Mathsnet

A variety of resources, including puzzles, tutorials, numeracy information and links to other maths websites.

www.nrich.maths.org.uk

Online Maths Club

Material for all ages and online answering service.

Science

www.wnet.org/savageearth

Savage Earth

Includes good introductions to plate tectonics, volcanoes, earthquakes and tsunamis (tidal waves). It has Flash animations and QuickTime movies and gives a useful overview without too many technicalities. Ideal for GCSE Science and useful for geology students.

www.bbc.co.uk/the_net/

BBC The Net

A magazine-style website on digital culture, including articles about the Internet and the issues surround it. Also provides links to some good web projects.

www.howstuffworks.com

How Stuff Works

Readable information to explain everyday technology and natural processes (such as car engines, combination locks, food, the immune system).

www.colorado.edu/physics/2000

Physics 2000

A journey through modern physics and its applications, with lots of interactive demonstrations and exercises. Materials geared to all levels and many wider topics, including health and food.

gslc.genetics.utah.edu

Genetic Science Learning Centre

This website provides genetic experiments which show how DNA, chromosomes and genes relate to conservation and forensic applications.

www.scicentral.com

SciCentral

Professional website which feeds news science stories to newspapers, TV and the Internet. See it here first. Includes an area designed to encourage science awareness among students.

Modern Languages

www.locuta.com

Centro Studi Italiani

A website specialising in Italian language learning. Has an electronic classroom area.

www.nene.ac.uk/lrs

Language resources

A University College Northampton website offering language resources for French, German, Italian and Spanish. Includes language learning, soundbites, quizzes, newspapers and links to other websites.

www.well.com

World Language Pages

A Liverpool John Moores University website offering a guide to language resources for many modern languages. It includes links to materials for language learning and suggestions about how best to use them.

Business

www.osl-ltd.co.uk

Oxford School of Learning

This website is designed for Business Studies at A level and has daily comprehension questions and essay plans on a wide range of topics. It has a number of Applied Business Studies questions too.

www.ft.com

Financial Times

The electronic version of the business newspaper. The website includes share prices, trends and currency conversion

www.economist.com

Economist

An online format of the popular business magazine which contains latest business news along with ideas, features, opinion and an analysis of international affairs.

www.oanda.com

Currency conversion

This website converts currency values and also contains a list of exchange rates for every day since 1990. Good for checking what your holiday money is worth before you go abroad and an excellent practical demonstration of currency conversion for maths students.

www.bized.ac.uk

Biz/ed

A website dedicated to business education with a huge pool of resources organised for UK qualifications and exams. Information includes data on 500 major companies, worksheets on the housing market, a virtual factor where A level and GNVQ business students can apply theory to a real-world business case.

www.bbc.co.uk/education/alvin/alvin.shtml

Investing for all

A BBC website designed for beginners. It contains FAQs (frequently asked questions), a 'jargon-buster' and a breakdown of what financial numbers mean.

Biology • Built Environment • Geography • Land and Environment • Leisure and Tourism

www.unfccc.de

Climate change

A United Nations website with useful beginner's guide to climate change, global warming and the ozone layer problem. The rest of the website contains in-depth information on all aspects of climate change.

www.nationalgeographic.com

National Geographic

Includes many interesting areas; a 'map machine' provides a choice of maps and views of anywhere in the world. Also includes information and tutorials on GIS.

www.scotese.com/earth.htm

Paleomap Project

A website devoted to the history of planet Earth, including detailed sequences of maps showing Earth over geological timescales. Good material for students studying continental drift in geography.

response.restoration.noaa.gov/kids/kids.htm

Environmental disasters

A US government website about the effects of oil spills or hazardous chemical accidents. Includes a guided tour to see how the experts deal with such environmental disasters, and simple experiments to investigate oil pollution and its effect on birds and animals.

www.countryside.gov.uk/what/f_forest.htm

Countryside agency

The website focuses on the new woodlands, community forests, on the edge of twelve major towns and cities. There is also information on National Parks, Areas of Outstanding Natural Beauty and walking trails for the countryside enthusiast.

www.terraquest.com

Virtual Antarctica

Use a virtual console to view images of Antarctica or learn about the ecology and wildlife of the region. Discover the history of Antarctic exploration and the environmental issues of the present.

www.ordsvy.gov.uk

The Ordnance Survey

A powerful geography resource for anything from local studies to contrasting localities and map-reading. The section on education includes aerial photography and wall maps.

www.schoolnet.ca/vp-pv/learning

Learning for a Sustainable Future

Activities cover air, biodiversity, habitat and infrastructure, production and consumption systems, social systems, soil, and water.

www.greatestplaces.org

The Greatest Places on Earth

Allows you to explore seven geographically different regions of the world. Includes sample video clips and soundbites which offer an insight into the culture and customs of the various countries.

www.the-education-site.com/emenu.html

Electricity

Contains information covering all of the main commercial power generation systems, including gas, coal, nuclear and wind.

Construction • Design and Technology • Engineering • Manufacturing

www.howstuffworks.com

How Stuff Works

Readable information to explain everyday technology and natural processes (such as car engines, combination locks, food, the immune system).

www.mmeade.com/cheat

Nutrition Cheat Sheet

A general interest website about food and health with information about vitamins, minerals and trace elements. There are summaries of what they do what happens if you don't get enough or too much, and recommended daily allowance (RDA).

www.innerauto.com

Autotour

A US website which examines the auto industry. You can see installation on videos and pictures and you can play an interactive game that lets you build your own car company.

www.ford.com

Ford Motor Company

The official website of the Ford Motor Company with information on its history and issues such as environmental policy. In addition to the specifications of current cars there are concept cars.

www.olen.com/food

Fast Food Facts

An interactive database that gives the nutritional values of different foods. The results include amounts of calories, fat, cholesterol, etc.

www.innerauto.com

Automotive Learning Online

This website has adopted a relatively simple approach to how a car and its moving parts work. The accompanying Java applets, along with clear descriptions of the function of individual components, make this an excellent starting point for the budding auto engineer, and for science and design and technology students.

Health and Social Care • Home Economics • Hospitality and Catering

www.nutrition.org.uk
British Nutrition Foundation
Food and nutrition information for consumers, design and technology students and teachers. There are useful GCSE and A level pages and a good list of other resources.

www.cre.gov.uk
Commission for Racial Equality
An excellent website on the work of the Campaign for Racial Equality. It covers racism and the law and it gives a list of FAQs.

www.bbc.co.uk/education/health/parenting/index.shtml
BBC Education – parenting
Contains common-sense advice for new parents on pregnancy, birth and child development. There are also facts on broader issues such as smoking during pregnancy and cot death.

www.open.gov.uk/doh/
Health of the Nation
A website from the Department of Health that covers national trends, including statistical data such as the percentage of children who smoke and the number of deaths through different types of cancer. A good source of health data with downloadable graphs, it includes details of national health targets.

www.olen.com/food
Fast Food Facts
An interactive database that gives the nutritional values of different foods. The results include amounts of calories, fat, cholesterol, etc.

www.thinkfast.co.uk/home.html
Thinkfast food website
Sponsored by the Health Education Authority, this website asks how can fast food be healthy? There is a quiz to test your knowledge and help you learn about the real properties of fast food and what it does to your body.

www.healthcalc.net
Healthcalc network
This website allows you to enter key data about your exercise level, diet, medical history and general health over a secure server, and find out how you shape up compared to the United States average.

www.wiredforhealth.gov.uk
Wired for Health
A government website which provides tailored health information and

appropriate links to other websites. Topics include sun safety, smoking, physical activity, healthy eating, alcohol, accidents and mental health.

www.ama-assn.org
Atlas of the body
Includes good descriptions of the body and systems such as the respiratory system, the skeleton, the reproductive system and the endocrine system.

Index